Age-appropriate Driver Assistance Systems

Heike Proff · Matthias Brand ·
Dieter Schramm
Editors

Age-appropriate Driver Assistance Systems

Technical, Psychological and Business Aspects

Springer

Editors
Prof. Dr. Heike Proff
Lehrstuhl für ABWL & Internationales Automobilmanagement
Universität Duisburg-Essen
Duisburg Nordrhein-Westfalen, Germany

Prof. Dr. Matthias Brand
Lehrstuhl für Allgemeine Psychologie: Kognition, Universität Duisburg-Essen
Duisburg, Nordrhein-Westfalen
Germany

Prof. Dr. Dieter Schramm
Lehrstuhl für Mechatronik, Universität Duisburg Essen
Duisburg Nordrhein-Westfalen, Germany

ISBN 978-3-658-44837-0 ISBN 978-3-658-44838-7 (eBook)
https://doi.org/10.1007/978-3-658-44838-7

Translation from the German language edition: "Altersgerechte Fahrerassistenzsysteme" by Heike Proff et al., © Der/die Herausgeber bzw. der/die Autor(en), exklusiv lizenziert durch Springer Fachmedien Wiesbaden GmbH, ein Teil von Springer Nature 2020. Published by Springer Fachmedien Wiesbaden. All Rights Reserved.

This book is a translation of the original German edition "Altersgerechte Fahrerassistenzsysteme" by Heike Proff, published by Springer Fachmedien Wiesbaden GmbH in 2020. The translation was done with the help of an artificial intelligence machine translation tool. A subsequent human revision was done primarily in terms of content, so that the book will read stylistically differently from a conventional translation. Springer Nature works continuously to further the development of tools for the production of books and on the related technologies to support the authors.

This Springer imprint is published by the registered company Springer Fachmedien Wiesbaden GmbH, part of Springer Nature.
The registered company address is: Abraham-Lincoln-Str. 46, 65189 Wiesbaden, Germany

Preface (Editor)

Age-appropriate driver assistance systems are becoming increasingly important in the aging societies of highly developed industrial countries with an increasing number of older drivers, in order to at least partially compensate for the age-related decline in cognitive and physical performance and thus increase the safety of all road users.

Although older people need support when driving and also have the purchasing power, thus potentially being a very receptive "Silver Market", the demand for driver assistance systems falls far short of the expectations of car manufacturers and suppliers. This may also be due to the fact that the offered driver assistance systems are not yet sufficiently tailored to the needs of this age group. Therefore, in the spring of 2017, the ALFASY project (**Al**tersgerechte **Fa**hrerassistenz**sy**steme, i.e. age-appropriate driver assistence systems) was started with a duration of three years. The aim of the project is the development, prototypical construction, and testing of an acoustic driver assistance system that is tailored to the needs of older drivers (50 years and older) and can be economically sensibly included in the product portfolio of car manufacturers and suppliers in an economically viable way.

For this, technical, psychological, and business studies were necessary. In the ALFASY project, the Chairs for Mechatronics, General Psychology: Cognition, and General Business Administration & International Automotive Management of the University of Duisburg-Essen worked together with Ford-Werke GmbH, HEAD acoustics GmbH, and Allround Team GmbH to test, improve, and retest previously offered driver assistance systems, both in the driving simulator and in real driving operations on a test track.

We are very pleased to be able to publish important results of the different work packages in this book. They would not have been possible without the financial support from the European Fund for Regional Development (EFRE) of the state of NRW and the EU, as well as from the PtJ in Jülich. The constant willingness to discuss of our practical partners, especially Dr. Stefan Wolter (Ford-Werke GmbH), Dr.-Ing. Magnus Schäfer (HEAD acoustics GmbH) and Martin Brockmann (Allround Team GmbH), as well as the tireless efforts of our employees Dr. Timo Günthner, Dr. Magnus Liebherr, Stephan Schweig and Melanie Zerr, to whom we are very grateful, were indispensable and extremely helpful. Finally, we would also like to thank Dr. Josip Jovic for the great support with all organizational questions of the project and the SpringerGabler publishing house, especially *Angela Schulze-Thomin and Nadine Teresa* who have supported the creation of the book.

We hope that the book offers suggestions, appeals to both scientists and practitioners, and encourages collaboration between different disciplines.

Duisburg
in July 2020

Heike Proff
Matthias Brand
Dieter Schramm

Contents

Editors and Contributors

About the Editors

Prof. Dr. Heike Proff, Studied business administration in Frankfurt and Mannheim, received a doctorate in Frankfurt, habilitation in Mannheim, research stays in Japan, Ghana, Korea, and the USA. From 2004 to 2009, held the Zeppelin Chair for International Management at Zeppelin University in Friedrichshafen. Since 2009, Chair of General Business Administration & International Automotive Management at the University of Duisburg-Essen, coordinator of the master's program "Automotive Engineering & Mobility Management" and organizer of the annual "Science Forum Mobility". Research focuses are Strategic and International Management, especially in the automotive industry, empirical studies mainly on the automotive industry, member of the "Program on Vehicle and Mobility Innovations (PVMI)" and in "The International Network of the Automobile (Gerpisa).

Prof. Dr. Matthias Brand, Born in 1975, he received his diploma in Psychology in 1999 from the University of Koblenz-Landau, earned his doctorate in 2001 and habilitation in 2005 from the University of Bielefeld. Since 2009, he has been a Professor of General Psychology: Cognition at the University of Duisburg-Essen and Director of the Erwin L. Hahn Institute for Magnetic Resonance Imaging in Essen. He is also the scientific director of the Center for Behavioral Addiction Research (CeBAR) at the University of Duisburg-Essen. For many years, he has been dealing with questions of human decision-making behavior, especially in the context of disorders caused by addictive behavior.

Prof. Dr.-Ing. Dieter Schramm, Studied mathematics and received a doctorate in 1986 from the University of Stuttgart, then worked for 18 years in the automotive industry at the companies Bosch and Tyco Electronic, among other things as a development manager and managing director. Since 2004, he has held the Chair of Mechatronics at the University of Duisburg-Essen, and from 2006 until 2023 2006, he was the Dean of the Faculty of Engineering.

Contributors

Prof. Dr. Matthias Brand, General Psychology: Cognition, University of Duisburg-Essen, Duisburg, Germany

Martin Brockmann, Allround Team GmbH, Cologne, Germany

Abhinav Dhake, Ford Works GmbH, Aachen, Germany

Dr. Timo Günthner, Chair of Business Administration & International Automotive Management, University of Duisburg-Essen, Duisburg, Germany

Pia Immoor, Allround Team GmbH, Erftstadt, Germany

Dr. Josip Jovic, Chair of Business Administration & International Automotive Management, University of Duisburg-Essen, Duisburg, Germany

Dr. Magnus Liebherr, University of Duisburg-Essen, Duisburg, Germany

Prof. Dr. Heike Proff, Chair of Business Administration & International Automotive Management, University of Duisburg-Essen, Duisburg, Germany

Prof. Dr.-Ing. Dieter Schramm, Chair of Mechatronics, University of Duisburg-Essen, Duisburg, Germany

Milan Schreiber, Allround Team GmbH, Cologne, Germany

Stephan Schweig, Chair of Mechatronics, University of Duisburg-Essen, Duisburg, Germany

Dr.-Ing. Magnus Schäfer, HEAD acoustics GmbH, Herzogenrath, Germany

Dr.-Ing. Carsten Starke, Ford Works GmbH, Aachen, Germany

Sophia Wingen, Allround Team GmbH, Cologne, Germany

Dr. Stefan Wolter, Ford Works GmbH, Aachen, Germany

Melanie Zerr, University of Duisburg-Essen, Duisburg, Germany

Lukas Zeymer, Chair of Business Administration & International Automotive Management, University of Duisburg-Essen, Duisburg, Germany

Mobility in Old Age—an Introduction

Heike Proff, Matthias Brand and Dieter Schramm

Contents

1 Problem Statement

The demographic change is becoming increasingly apparent, especially in industrialized countries. According to a current projection by the Federal Statistical Office, the proportion of older people (50 years and older) will increase from

Prof. Dr. Heike Proff, Prof. Dr. Matthias Brand, Prof. Dr.-Ing. Dieter Schramm, all University of Duisburg-Essen.

H. Proff (✉)
Lehrstuhl für ABWL & Internationales Automobilmanagement,
Universität Duisburg-Essen, Duisburg, Germany
e-mail: heike.proff@uni-due.de

M. Brand
Allgemeine Psychologie: Kognition, Universität Duisburg-Essen, Duisburg, Germany
e-mail: matthias.brand@uni-due.de

D. Schramm
Lehrstuhl für Mechatronik, Universität Duisburg-Essen, Duisburg, Germany
e-mail: dieter.schramm@uni-due.de

H. Proff et al. (eds.), *Age-appropriate Driver Assistance Systems*,
https://doi.org/10.1007/978-3-658-44838-7_1

40% in 2009 to an estimated 46% in 2030 and up to 49% in 2060 (Statisches Bundesamt 2019).

Older people are mobile, as mobile as never before (Schlag, 2013). About 58 million people in Germany have a driving license, 43% of them are 50 years and older (Federal Motor Transport Authority, 2018). For many older people, a car facilitates independence, autonomy, and social participation (Whelan et al., 2006), which in turn is associated with an increase in quality of life (Engeln, 2001; Burghard, 2005).

Older drivers tend to have more problems in traffic than younger ones, especially in complex traffic situations such as turning left against oncoming traffic, merging and changing lanes on highways at high speed (Chandraratna & Stamatiadis, 2003). In addition, older road users show increased impairments due to visual changes such as fog, rain, or twilight (e.g., McGwin Jr. et al., 2000; Gruber et al., 2013). Besides physiological changes (e.g., deterioration of visual ability), neuropsychological changes within cognitive domains play a major role in explaining age-associated limitations in driving performance. In the aging process of a healthy person, a number of cognitive abilities are preserved, such as crystalline intelligence, i.e., semantic knowledge and "age wisdom", but specific cognitive functions experience an age-associated performance decline (Brand & Markowitsch, 2010; Markowitsch et al., 2005). These particularly affect memory performance and so-called executive functions such as the targeted suppression of action tendencies, cognitive flexibility, planning, working memory, and the selection of strategies, which in turn are associated with successful driving (Reuter-Lorenz & Sylvester, 2005; Uekermann et al., 2006; Ashendorf & McCaffrey, 2008; Hodzik & Lemaire, 2011). At the brain level, age-associated structural and functional declines are (initially) primarily seen in the frontal brain and hippocampus formation, which are considered neural correlates of cognitive performance losses in old age (Fiell & Walhovd, 2010).

In addition to an age-related decline in cognitive performance, visual abilities also typically decrease, which can significantly impair vehicle operation (Davidse, 2007). This affects less the pure visual acuity, which is usually sufficient in older drivers and whose decline usually becomes critical only at a very advanced age, but rather the dynamic visual performance, the detection of movements, night vision ability, the width of the field of vision, and the glare and contrast sensitivity (Fisk et al., 2009). Motor skills also decrease with age. A poorer joint mobility, such as a weakened neck mobility, results in a restricted field of vision in older drivers (MoPact, 2014).

Thus, there is a group of older drivers who still actively participate in road traffic and whose cognitive and/or physiological performances are so severely reduced that

Fig. 1 Task of the project: Increasing the objective safety in road traffic for all road users through suitable assistance systems. *Source* own design, drawing by Pedro Ribeiro Ferreira

safe driving is endangered (cf. ibid.). Therefore, it is the task of science and industry to increase the objective safety and independence in road traffic for all road users through suitable technical solutions and assistance systems (Fig. 1). These systems enable a compensation of the impaired information intake and processing compared to younger drivers and ensure active participation in motorized individual traffic for older people, taking into account the need for safety (Kubitzki, 2013).

Technologies are already available; there are advanced driver assistance systems that help both older and younger drivers (Wild, 2014) and are constantly being improved. Such driver assistance systems include semi-automated systems, such as speed and distance control systems or Adaptive Cruise Control (ACC), lane keeping assist, emergency braking and warning systems e.g. for obstacles and parking aids. They are seen as steps of the digital transformation towards autonomous driving (i.e. up to level 5 "no driver") and today enable level 2 (Fig. 2).

Older people are consistently capable and predominantly also willing to pay a premium for products that increase safety (Wild, 2014). They belong to the most affluent population groups. According to federal and state studies and investigations, their financial living conditions are relatively favorable. Table 1 differentiates the monthly household net income by age groups.

Not only income, but also significant life events influence purchasing behavior, so that age is not likely to have a linear effect on the needs and demand for driver assistance systems. This is shown by life course studies of critical life events (see e.g. Hjelmar, 2011). These usually include age-dependent events such

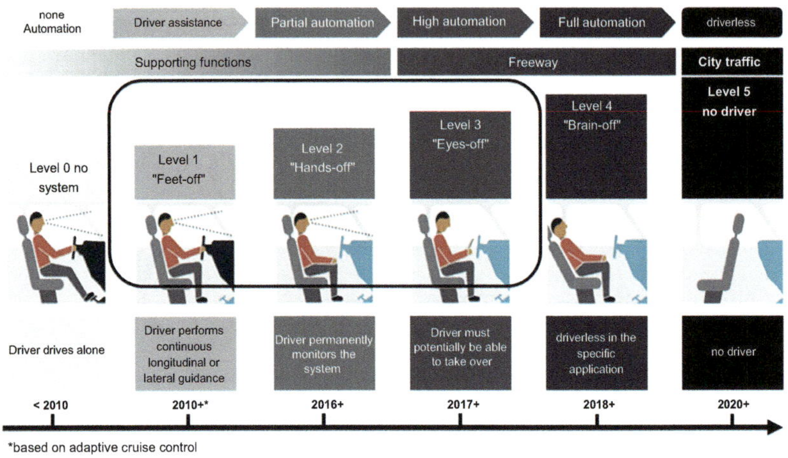

Fig. 2 Driver assistance systems on the way to autonomous driving. *Source* according to VDA, 2015

Table 1 Monthly net income in different age groups (in million). *Source* Statistisches Bundesamt, 2018

		Population (in millions in Germany)	Average Monthly Net income per Budget in euros (2017)
Age groups	55-65	11.0	3,317
	65-70	4.0	2,669
	70-80	8.5	2,534
	über 80	4.5	2,367

as studying, starting a career, starting a family and ending professional activity, but also non-age-dependent events such as unemployment, life-threatening illnesses, the death of relatives or the partner (e.g. Elder, 1994; Müggenburg et al., 2015; Uteng et al., 2019). Initial studies confirm the assumption that age, but also critical life events in old age, influence willingness to pay. King, Jr. et al., (2005) demonstrate, for example, that older people with increasing income are more willing to pay more for better mobility.

Although many older people need driver assistance systems, especially when critical life events increase the need and purchasing power is available, demand for driver assistance systems on the potentially very receptive "Silver Market" (e.g. Kohlbacher & Herstatt, 2011; Klimczuk, 2016) lags far behind the expectations of manufacturers and suppliers (Winner & Schopper, 2015).

This may possibly be explained by the fact that the built-in and offered driver assistance systems are still too little oriented towards the needs of older people. But it could also be that those affected are not sufficiently informed about how they can be helped. Since systematic investigations are lacking as to which driver assistance systems need to be developed so that older people accept and use them and traffic safety is improved (Pelizäus-Hoffmeister, 2013; Hartwich, 2017), age-appropriate assistance systems are so far hardly explicitly and aggressively offered by car manufacturers and suppliers. They are often not even mentioned or highlighted in special equipment.

It is therefore necessary to develop age-appropriate vehicle and safety concepts to avoid accidents and impairments and to test their suitability for older drivers, who often have only a relatively low affinity for technology and react more sensitively to poorly defined and implemented systems than younger ones (Davidse, 2007).

2 Objective and Course of the Investigation

In view of the so far low demand for driver assistance systems by older people, the ALFASY project (**<u>Al</u>**tersgerechte **Fa**hrerassistenzsysteme, i.e. age-appropriate driver assistence systems) was started in the spring of 2017 with the aim of developing, building and testing an acoustic human-machine interface for a selection of assistance systems that is tailored to the needs of the steadily growing group of older drivers (50 years and older) and can be included in the product portfolio of car manufacturers and suppliers in an economically viable way. To achieve this goal, three chairs of the University of Duisburg-Essen (Chair of Mechatronics, Chair of General Psychology: Cognition and Chair of General Business Administration & International Automotive Management), a car manufacturer (Ford-Werke GmbH), a supplier (HEAD acoustics GmbH) and a service provider (Allround Team GmbH) worked together in the ALFASY project.

Important questions in the development of driver assistance systems are when they are activated, how they interact with the driver, how communication between the driver and the driver assistance system (acoustically, visually or haptically) is carried out and also how the attention of older drivers is aroused without overwhelming them (Rudinger et al., 2013). Because driver assistance systems can also have negative effects if drivers rely too much on the takeover of functions by assistance systems or if there is a "sensory overload" especially for older drivers, learning opportunities are demanded in dealing with modern vehicle technologies that are also aimed at older drivers and take into account the skills and cognitive peculiarities of all age groups (EU, 2015). This allows display and operating systems to be designed in an age-appropriate manner. This leads to an increase in safety in road traffic, as stress is reduced and the driver can concentrate on the actual traffic situation.

In the ALFASY project, driver assistance systems are first examined as to how signals, stimuli and hints are perceived, processed and used for the adaptation of driving behavior, so that the driver can make decisions very quickly with the available information about the traffic situation. This requires cognitive executive functions, which generally and especially in the pathological aging process experience significant losses. The compensation of these losses is to be achieved by the assistance system to be developed.

On the basis of empirical data, a framework booklet and related models that show or demonstrate the feasibility of driver assistance systems (demonstrators) as well as a prototype of a driver assistance system with an acoustic human-machine interface for older drivers with cognitive and physical performance impairments should then be developed for the first time. It was tested in a driving simulator and on a real test track with a larger sample of older drivers from different age groups. To this end, warning or hint signals had to be translated into acoustic stimuli in dealing with potentially critical driving situations that meet the needs of the target group. Demonstrators and prototype should support older drivers in the time between cognition and decision through an interface and overcompensate for additional costs and greater complexity due to the large number of variants through higher revenues. Economic, psychological, design and technical implications for business and science are to be derived and exploited from the research results. Figure 3 provides an overview of the project structure.

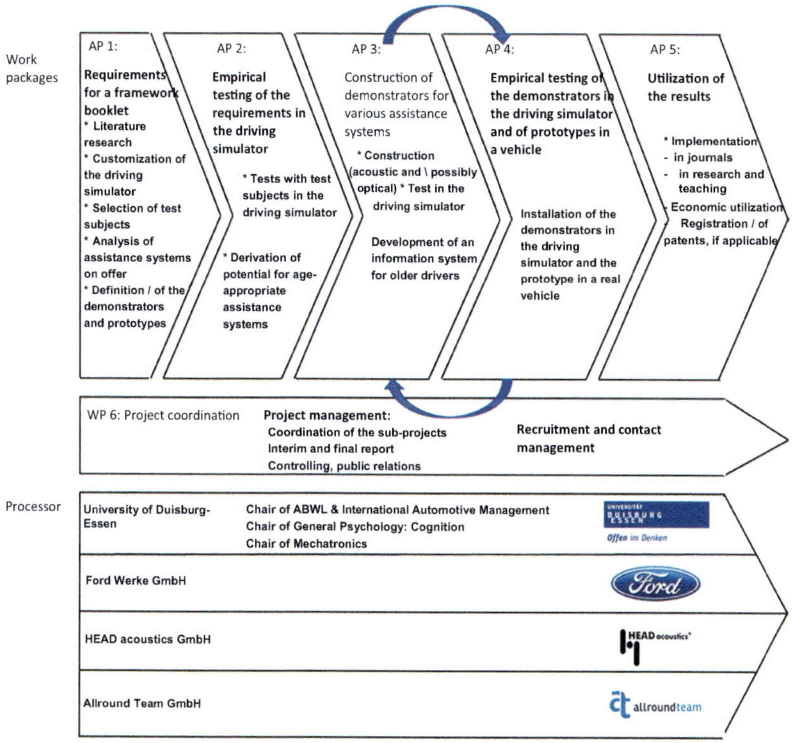

Fig. 3 Structure of the ALFASY project. *Source* own design

3 Structure of the Book

This book provides information on important project results of the different work packages and offers a summary of the results.

Chapter "*Older Car Drivers: Investigation and Sample in the ALFASY Project*" (Jovic, Günthner, Proff, Chair of General Business Administration & International Automotive Management) presents the panel of the investigation of older car drivers: a multi-stage investigation design with

- a preliminary investigation of the multi-stage driving suitability of the test subjects, their demographic characteristics (especially age), their latent and open needs (age-related restrictions and support potentials) as well as their concrete needs (e.g., experiences and customer wishes for driver assistance systems depending on age),
- two main investigations in the driving simulator, in which neurophysiological reactions, driving behavior as well as acceptance and willingness to pay are first examined for already offered driver assistance systems and then for improved (acoustic) systems, and
- a third investigation of further improved acoustic assistance systems with a smaller, somewhat differently composed sample in a vehicle on a test track.

The Chapter "Driver Assistance Systems—An Overview" (Schramm, Schweig, Chair of Mechatronics) provides an overview of driver assistance systems and their future development. It describes the assistance systems used in the ALFASY project (cross-traffic assistant, parking aid, lane keeping assistant, blind spot assistant, and tire pressure monitoring system) and their functionality. They were selected to make them perceptible to older people using acoustic signals and to scientifically investigate the benefits of these systems and the interface between humans and machines for this population group.

The Chapter "Driver Assistance Systems in the Context of Age-Appropriate HMI Design" (Brockmann, Schreiber, Wingen, Immoor, Allround Team GmbH) considers driver assistance systems in the context of an age-appropriate design of the interface between humans and machines (Human-Machine Interface, HMI), because poor usability and understandability of the user interface can be a reason why the proportion of older drivers who actually own and use driver assistance systems is very low (e.g., Trübswetter, 2015). By evaluating studies, it is to be found out to what extent already available assistance systems benefit older people and what problems they have due to age-related deficits in driving and interacting with user interfaces. Based on this analysis, assistance systems and interfaces can be designed that take into account the needs of this group of people (Kupschick et al., 2019).

The Chapter "Evaluation of the Performance of Acoustic Assistance Systems for Older Drivers" specifically evaluates the performance of acoustic assistance systems for older people (Schäfer, HEAD acoustics GmbH), because the reaction speed to acoustic stimuli is higher than to visual and haptic stimuli (e.g., Aditya et al., 2015). It must be an important optimization goal of all assistance systems to convey time-critical information as clearly as possible. Therefore,

- first a brief overview of the localization of sound sources from a technical perspective and from human perception is provided.
- Then a new microphone array (interaction of several microphones) is introduced, which allows to provide technical methods exactly the components of a sound field that a human listener also uses for localization.
- Finally, a method is described and evaluated in terms of its performance, which allows localization of sound sources using machine learning methods and can also be adopted for other use cases or user groups.

The Chapter "Driving Behavior of Older People in the Driving Simulator" deals with the driving behavior of older drivers in the driving simulator (Schweig, Schramm, Chair of Mechatronics):

- First, the structure of the simulator and the experiments carried out are described and the test results are briefly summarized.
- It is then shown that the great challenge of all studies in a simulator is the frequently occurring simulator sickness (kinetosis). This disease is attributed to the human body's lack of adaptation to unfamiliar combinations of sensory stimuli and thus also as a difference, for example, between optical stimuli and acceleration. It can cause headaches, sweat, dry mouth, drowsiness, disorientation, dizziness, nausea or/and vomiting (Brooks et al. 2010) and lead to the termination of simulation attempts, more so among women than among men.
- Finally, the Index of Performance (IOP) (Joshi et al., 2017) is calculated for an objective evaluation of driving performance.

In the ALFASY project, following the investigations in the driving simulator, the third main investigation was carried out in a real vehicle (Wolter, Dhake, Starke, Ford-Werke GmbH, Chapter "Implementation of the Real Driving Studies—Test Vehicle and Experimental Setup in t2"). The aim of this investigation was to test the individual (acoustic) driver assistance systems under real driving conditions. For this purpose, a test vehicle was appropriately converted.

Chapter "Neuropsychological and -physiological correlates of driving behavior of older drivers within simulated environments" examines neuropsychological and -physiological correlates of driving behavior of older drivers within simulated environments (Liebherr, Zerr, Brand, Chair of General Psychology: Cognition). Driving simulators offer a good opportunity to investigate potential influence and interaction effects on driving performance (highly standardized, cost- and time-efficient, and free from accidents or injuries). However, due to the differences between real and simulated driving, problems often arise such as simulator sickness and inadequate adaptation to the simulator (Brooks et al., 2010), which can affect the driving performance of the subjects. These aspects are addressed based on three studies and results on the topics

- cognitive abilities, simulator adaptation or reaction, and simulator sickness,
- stress reactions in the simulator, and
- attention and driving behavior

are described, each taking into account the age of the subjects.

Chapter "Market potentials of older drivers - willingness to pay and acceptance of age-appropriate driver assistance systems" (Günthner, Zeymer, Proff and Jovic, Chair of General Business Administration & International Automotive Management) summarizes the results of the study on acceptance and willingness to pay of older people. For this purpose, important results of previous research are first presented, but also explanations for the willingness to pay and the acceptance of driver assistance systems for older people as well as for the influence of critical life events are sought. Then the study concept and study results of the first main study on willingness to pay for the offered driver assistance systems and on acceptance are presented and compared with the results of the second main study on improved acoustic driver assistance systems. The results of the third main study after a drive on a test track are only briefly mentioned, because the sample is relatively small and hardly comparable with the previous studies. From this, deeper insights into the market potential of older drivers can be derived.

In Chapter: "Age-appropriate driver assistance systems:Technical, psychological and business aspects—A summary", the editors Proff, Brand and Schramm summarize important findings from the ALFASY project.

References

Aditya, J., Bansal, R., Kumar, A. & Singh, K. (2015). A comparative study of visual and auditory reaction times on the basis of gender and physical activity levels of medical first year students. *International Journal of Applied and Basic Medical Research,* 5(2), 124–127.

Ashendorf, L., & McCaffrey, R. J. (2008). Exploring age-related decline on the Wisconsin Card Sorting Test. *Clinical Neuropsychologist.* 22, 262–272. https://doi.org/10.1080/13854040701218436.

Brand, M., Markowitsch, H.J. (2010). Aging and decision-making: A neurocognitive perspective. *Gerontology, 56*(3), 319–324.

Brooks, J.O., Goodenough, R R., Crisler, M.C., Klein, N.D., Alley, R.L., Koon, B.L., & Wills, R.F. (2010). Simulator sickness during driving simulation studies. *Accident Analysis & Prevention, 42*(3), 788–796.

Burghard, E. (2005). *Fahrkompetenz im Alter – Die Aussagekraft diagnostischer Instrumente bei Senioren und neurologischen Patienten.* München (=Dissertation, Ludwig-Maximilian-Universität München).

Chandraratna, S., & Stamatiadis, N. (2003). Problem driving maneuvers of elderly drivers. *Transportation Research Record, 1843*(1), 89–95.

Davidse, R. (2007): *Assisting the older driver. Intersection design and in-car devices to improve the safety of the older driver*. Leidschendam, NL: SWOV-Dissertatiereeks.

Elder, G. H. (1994). Time, human agency and social change: Perspectives on the life course. *Social psycology quarterly, 57*(1), 4–15.

Engeln, A. (2001). *Aktivität und Mobilität im Alternsprozess*. Aachen: Shaker.

EU (2015). *People in the EU: who are we and how do we live?* 2015 edition. Luxembourg: eurostat. Statistical Books.

Fiell, A. M., & Walhovd, K. B. (2010). Structural brain changes in aging: courses, causes and cognitive consequences. *Reviews in the Neuroscience, 21*, 187–221. https://doi.org/10.1515/revneuro.2010.21.3.187.

Fisk, A.D., Rogers, W.A., Charness, N., Czaja, S.J., &Sharit, J. (2009). *Designing for older adults: principles and creative human factors approaches*, 2. Auflage, Boca Raton: CRC Press.

Gruber, N., Mosimann, U. P., Müri, R. M., & Nef, T. (2013). Vision and night driving abilities of elderly drivers. *Traffic injury prevention, 14*(5), 477–485.

Hartwich, F. (2017). Supporting older drivers through emerging in-vehicle technologies: Performance-related aspects and user acceptance. (=Dissertation, Technische Universität Chemnitz).

Hjelmar, U. (2011). Consumers' purchase of organic food products. A matter of convenience and reflexive practices. *Appetite, 56*(2), 336–344.

Hodzik, S., & Lemaire, P. (2011). Inhibition and shifting capacities mediate adults' age-related differences in strategy selection and repertoire. *Acta Psychologica, 137*, 335–344. https://doi.org/10.1016/j.actpsy.2011.04.002.

Joshi, S. S., Maas, N., & Schramm, D. (2017). A vehicle dynamics based algorithm for driver evaluation. In *11th International Conference on intelligent Systems and Control (ISCO)*, S. 5, Karpagam College of Engineering Myleripalayam, Coimbatore, India.

King, J.T. Jr., Tsevat, J., Lave, J.R., & Roberts, M.S. (2005). Willingness to pay for a quality-adjusted life year: Implications for societal health care resource allocation. *Medical Decision Making, 25*(6), 667–677.

Klimczuk, A. (2016). Comparative analysis of national and regional models of the silver economy in the European Union. *International Journal of Ageing and Later Life, 10*(2), 31–59.

Kohlbacher, F., & Herstatt, C. (eds.). (2011). *The silver market phenomenon. Marketing and innovation in the aging society.* Berlin: Springer.

Kraftfahrt-Bundesamt (2018). Bestand an allgemeinen Fahrerlaubnissen im ZFER am 1. Januar 2018 nach Geschlecht, Lebensalter und Fahrerlaubnisklassen. (=Abgerufen am

10.03.2020 von https://www.kba.de/DE/Statistik/Kraftfahrer/Fahrerlaubnisse/Fahrerlaubnisbestand/2018/2018_fe_b_geschlecht_alter_fahrerlaubniskl.html?nn=2218648).

Kubitzki, J. (2013). *Sicherheit und Mobilität älterer Verkehrsteilnehmer*. Fachtagung zum Verkehrssicherheitsprogramm des Landes Brandenburg. (=Abgerufen am 10.03.2020 von https://www.iges.com/sites/iges.de/myzms/content/e2856/e4186/e4668/e8405/e8384/e8385/attr_objs8392/130307_IGES_Verkehrssicherheit_Workshop_3_Kubitzki_ger.pdf).

Kupschick, S., Bürgelen, J., Jürgensohn, T., & Protzak, J. (2019). *Erhöhung der Verkehrssicherheit älterer Kraftfahrer durch Verbesserung ihrer visuellen Aufmerksamkeit mittels "Sehfeldassistent"*. (=Berichte der Bundesanstalt für Straßenwesen. Unterreihe Fahrzeugtechnik, 127).

Markowitsch, H. J., Brand, M., & Reinkemeier, M. (2005). Neuropsychologische Aspekte des Alterns. In: S.-H. Filipp, U. M. Staudinger (eds.): *Entwicklungspsychologie des mittleren und höheren Erwachsenenalters*. Göttingen: Hogrefe, pp. 79–122.

McGwin Jr, G., Chapman, V., & Owsley, C. (2000). Visual risk factors for driving difficulty among older drivers. *Accident Analysis & Prevention, 32*(6), 735–744.

MoPact (2014). *Mobilising the potential of active ageing in Europe*. (= Abgerufen am 09.03.2020 von http://mopact.group.shef.ac.uk/)

Müggenburg, H., Busch-Geertsema, A., & Lanzendorf, M. (2015). Mobility biographies: A review of achievements and challenges of the mobility biographies approach and a framework for further research. *Journal of Transport Geography*, 46, 151–163.

Pelizäus-Hoffmeister, H. (2013). *Zur Bedeutung von Technik im Alltag Älterer. Theorie und Empirie aus soziologischer Perspektive*. Wiesbaden: Springer.

Reuter-Lorenz, P. A., & Sylvester, C.-Y. (2005). The cognitive neuroscience of working memory and aging. In: R. Cabeza, L. Nyberg, D. Park (eds.), *Cognitive Neuroscience of Aging*. Oxford: Oxford University Press.

Rudinger, G. (2013). *Ältere Verkehrsteilnehmer: Gefährdet oder gefährlich? Defizite, Kompensationsmechanismen und Präventionsmöglichkeiten*. (=Präsentation 18.4. 2013, Wissenschaftszentrum Bonn).

Schlag, B. (2013). Persönliche Veränderungen der Mobilität und der Leistungsfähigkeit im Alter. In: B. Schlag, K. J. Beckmann (eds.): *Mobilität und demografische Entwicklung*, Köln: TÜV Media, pp.119–143. (=Schriftenreihe Mobilität und Alter der Eugen-Otto-Butz-Stiftung, Bd. 7).

Statistisches Bundesamt (destatis) (2018). *Wirtschaftsrechnungen Laufende Wirtschaftsrechnungen Einkommen, Einnahmen und Ausgaben privater Haushalte*. Wiesbaden. (=Fachserie 15 Reihe 1).

Statistisches Bundesamt (2019). *Bevölkerung Deutschlands bis 2060: Ergebnisse der 14. koordinierten Bevölkerungsvorausberechnung – Hauptvarianten 1 bis 9*. Wiesbaden.

Trübswetter, N., & Bengler, K. (2013). *Why should I use ADAS? Advanced driver assistance systems and the elderly: Knowledge, experience and usage Barriers*. In: Proceedings of the 7th International Driving Symposium on Human Factors in Driver Assessment, Training, and Vehicle Design, University of Iowa, Iowa City, 495–501.

Uekermann, J., Channon, S., & Daum, I. (2006). Humor processing, mentalizing, and executive function in normal aging. *Journal of International Neuropsychological Society*, 12, 184–911. https://doi.org/10.1017/s1355617706060280

Uteng, T. P., Julsrud, T.E., & George, C. (2019). The role of life events and context in type of car share uptake: Comparing users of peer-to-peer and cooperative programs in Oslo, Norway. *Transportation Research Part D*, 71, 186–206.

VDA (Verband der Automobilindustrie) (2015). *Automatisierung – Von Fahrerassistenzsystemen zum automatisierten Fahren*. Berlin.

Whelan, M., Langford, J., Oxley, J., Koppel, S., & Charlton, J. (2006). *The elderly and mobility: A review of the literature*. (=Monash University, Accident Research Center, Report 225).

Wild, A. (2014). Zwischen Wunsch und Wirklichkeit: Fahrerassistenzsysteme für ältere Autofahrer. *Zeitschrift für die gesamte Wertschöpfungskette Automobil*, 17(1), 58–63.

Winner, H., & Schopper, M. (2015). Adaptive cruise control. In *H. Winner/S. Hakuli, G./ Wolf (eds.): Handbuch Fahrerassistenzsysteme – Grundlagen, Komponenten und Systeme für aktive Sicherheit und Komfort*. Wiesbaden, Vieweg+Teubner: 851–891.

Older Drivers: Investigation and Sample in the Alfasy Project

Josip Jovic, Timo Günthner and Heike Proff

Contents

In order to iteratively develop, build, and test an acoustic driver assistance system in the project "Age-appropriate Driver Assistance Systems" (ALFASY), which is tailored to the needs of the steadily growing group of older drivers (50 years and older) and can be included in the product portfolio of car manufacturers and

Dr. Josip Jovic, Dr. Timo Günthner, Prof. Dr. Heike Proff, all University of Duisburg-Essen.

J. Jovic · T. Günthner · H. Proff (✉)
Lehrstuhl für ABWL & Internationales Automobilmanagement,
Universität Duisburg-Essen, Duisburg, Germany
e-mail: heike.proff@uni-due.de

J. Jovic
e-mail: josip.jovic@uni-due.de

T. Günthner
e-mail: timo.guenthner@uni-due.de

H. Proff et al. (eds.), *Age-appropriate Driver Assistance Systems*,
https://doi.org/10.1007/978-3-658-44838-7_2

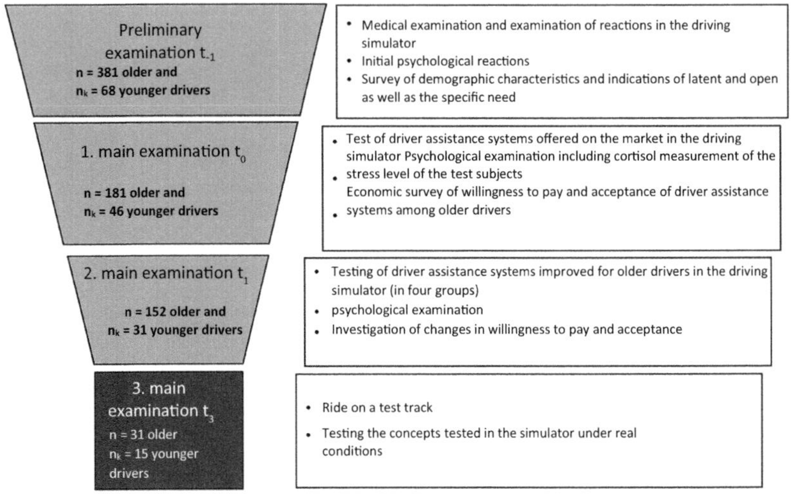

Fig. 1 Research design

suppliers, a broad empirical basis was important. Sufficient test subjects had to be recruited for the longitudinal study, which was divided into a preliminary study and three consecutive main studies (see Fig. 1).

This chapter describes, after a brief overview of the study design (Sect. 1), important characteristics of the first sample with 381 older drivers compared to 58 younger drivers of a comparison group, who participated in the preliminary examination. These are initiallydemographic characteristics such as age, gender, net income, and driving performance(Sect. 2). Further characteristics provid-einformation on latent and open needs (age-related restrictions and support potentials) aswell as on the specific need, product knowledge, experiences, expectations, and customerwishes for driver assistance systems depending on age (Sect. 3).Section 4 shows how the sample has changed in thesubsequent main examinations, before the study results are summarized in Sect. 5.

1 Research Design

In the ALFASY project, previously offered and improved acoustic driver assistance systems were to be tested among older drivers compared to a control group of younger drivers: initially multiple times in the driving simulator, finally in a vehicle on a test track.

Due to the sometimes high age of the subjects, a query of pre-existing conditions such as acute cardiovascular problems was necessary to exclude individuals with critical illnesses such as epilepsy, dementia, and Parkinson's disease from the tests. Blood pressure, pulse, and oxygen saturation were measured to get a picture of the health status. In the preliminary examination, reactions in the driving simulator (see Chapter "Driving Behavior of Older People in the Driving Simulator" on the driving behavior of older people) were tested, as empirical studies in the driving simulator had to reckon with the so-called "simulator sickness" , which can lead to headaches, dizziness, disorientation or nausea and thus to a termination of the tests, because the test subjects are deceived a movement, although they actually do not move. Studies show that older people are more often affected by "simulator sickness" than younger people (Brooks et al. 2010; Roenker et al. 2003; Schweig et al. 2018). In order for the planned 150 older drivers to participate in the first two main examinations in the simulator, 381 older test subjects were recruited for the longitudinal study (see Fig. 1).

In 181 of the 381 older test subjects and 46 of the 68 younger test subjects, driver assistance systems offered on the market were examined in the first main examination in mid-2018 in the driving simulator (see Chapter "Driver Assistance Systems—an Overview"), further psychological tests were carried out (see Chapter "Neuropsychological and -Physiological Correlates of the Driving Behavior of Older Drivers Within Simulated Environments"), and acceptance and willingness to pay were queried to estimate the market potentials of age-appropriate driver assistance systems (see Chapter "Market Potentials of Older Drivers - Willingness to Pay and Acceptance of Age-Appropriate Driver Assistance Systems").

151 of the older and 31 of the younger test subjects were also willing to participate in the second main examination at the end of 2018 and beginning of 2019. With them, acoustically improved driver assistance systems specifically for older drivers were tested (see Chapters "Driver Assistance Systems in the Context of Age-Appropriate HMI Design" and "Evaluation of the Performance of Acoustic Assistance Systems for Older DriversEvaluation"), especially the human-machine interface. The test subjects were divided into four groups to test different manifestations of the improved assistance systems. In addition, economic and psycho-

logical questions were asked (Chapters "Neuropsychological and -Physiological Correlates of the Driving Behavior of Older Drivers Within Simulated Environments" and "Market Potentials of Older Drivers - Willingness to Pay and Acceptance of Age-Appropriate Driver Assistance Systems").

In the third main examination in the fall of 2019, the acoustically improved assistance systems for the second main examination were tested with 31 older and 15 younger drivers in a vehicle on a test track (see Chapter "Implementation of the Real Driving Studies—Test Vehicle and Experimental Setup in t2"). Due to the spatial distance of the test track, not only test subjects of the preliminary examination could participate (see Sect. 4). This examination was also supplemented for the practice partners by selected economic questions. However, the results cannot be compared with those of the first two main examinations (Fig. 1).

2 Preliminary Examination—Demographic Characteristics of Older Test Subjects

In the preliminary examination (September 2017 to January 2018), demographic characteristics of the older test subjects were first determined: age, gender, educational attainment, and monthly net income, as well as average kilometers driven.

As already mentioned, 449 participants took part in this study, 381 older test subjects (50 years and older) and 68 younger test subjects. Figure 2 shows the age distribution of the sample. The older participants are on average 65 years old, 31% between 50 and 59 years old, 40% between 60 and 69 years, the rest 70 years and older, of which half are between 70 and 74 years. 17 participants are older than 80 years, the oldest participant is 92 years old. The average age of the younger comparison group is 29 years. Just under 60% are between 20 and 39 years old, the rest between 30 and 49 years.

In all age groups, the proportion of men is significantly higher. In the study group, two-thirds of the older test subjects were male, in the comparison group just under 60% (Fig. 3).

Of the 381 respondents in the study group, seven percent have a secondary school diploma, 13% have intermediate maturity, 20% have completed apprenticeship, 11% have a technical college entrance qualification, 10% have a high school diploma, 37% have a technical college/university degree, of which two percent have a doctorate. In the comparison group (younger drivers), significantly more have a high school diploma (38%) (Fig. 4).

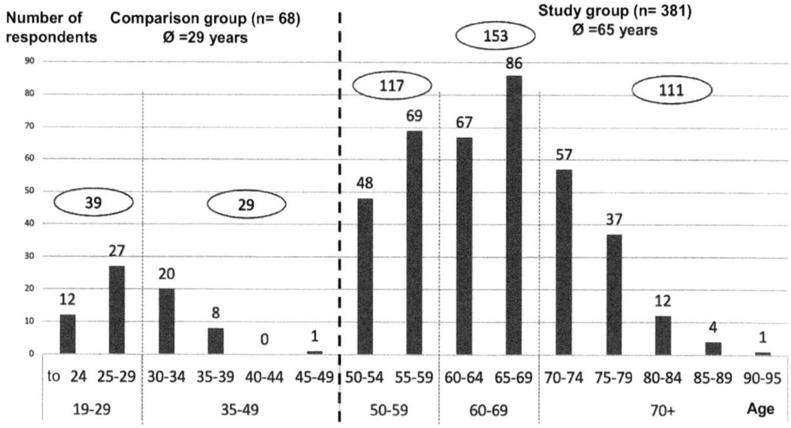

Fig. 2 Age distribution of the sample (n = 449)

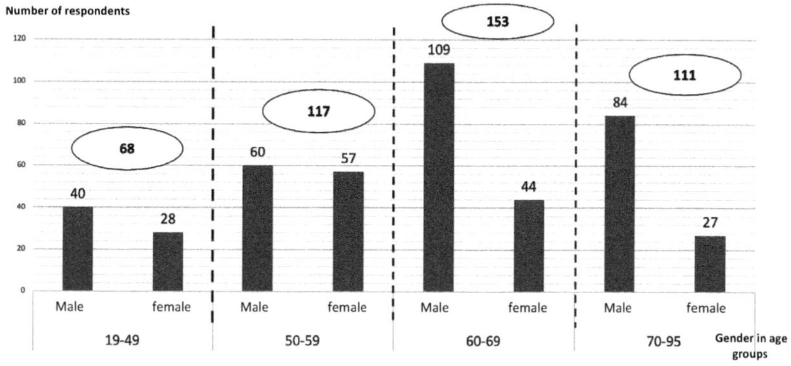

Fig. 3 Gender distribution by age groups in the preliminary examination (n = 449)

Figure 5 shows that the net income in the sample (n_{total} = 449) and of the over 50-year-old participants (n = 381) increases with age: from 1,000 to 2,000 EUR at the age between 20 and 49 years to 2,000 to 3,000 EUR in the age group between 50 and 59 years, to about 3,000 EUR in the age group of 60 to 69-year-olds and the even older.

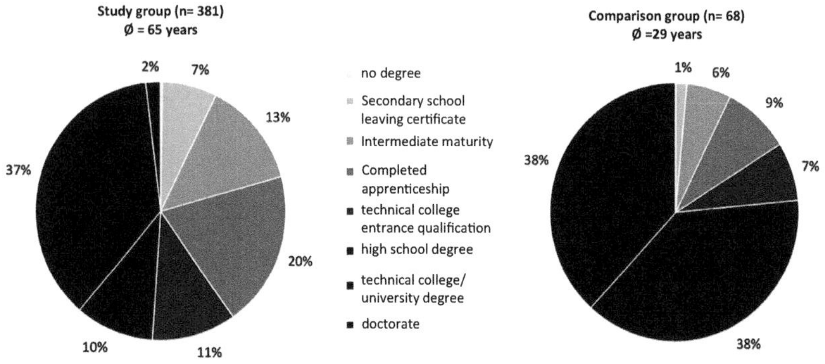

Fig. 4 Distribution of educational qualifications in the preliminary examination (n = 449)

The average driving performance of all respondents is between 21 and 50 km per day. The 50- to 59-year-olds drive slightly more than respondents of other age groups. With increasing age, the daily driven kilometers decrease (see Fig. 6).

3 Indications of Open and Latent Needs of the Test Subjects

In the Chapter "Mobility in Old Age—An Introduction" it was shown that older people need support when driving and generally have the purchasing power for driver assistance systems. Critical life events, such as the death of a life partner, can increase interest in driver assistance systems. These age groups should be a very promising "Silver Market" in at least more developed countries (Kohlbacher and Herstatt 2011). However, the sales of driver assistance systems still fall far short of the expectations of car manufacturers (Winner & Schopper 2015). Three reasons are given for this:

- older drivers do not miss driver assistance systems (no subjectively perceived lack that they want to eliminate, Homburg 2014),
- older drivers do have a (latent) need for assistance systems, but they are not (yet) aware of it, and
- older drivers do feel a (open) need, but not yet a demand, which is associated with the willingness to spend money to eliminate it (Homburg and Krohmer 2006).

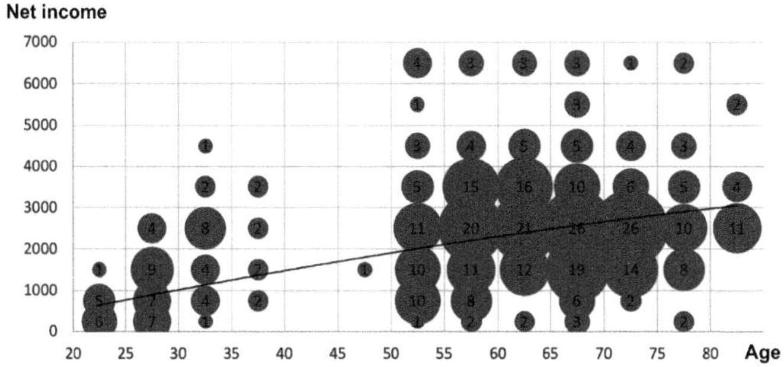

Fig. 5 Monthly net income by age in the preliminary examination (n = 449)

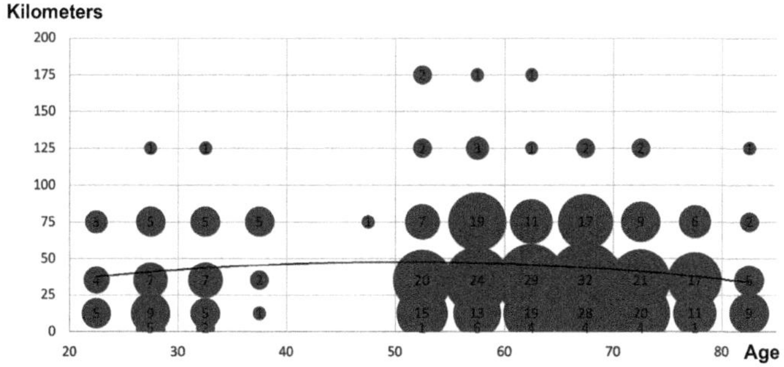

Fig. 6 Average kilometers driven by age in the preliminary examination (n = 449)

To gain insights into this, both

1. latent needs (deficits that are not yet perceived) about the effects of age-related physical restrictions were asked in the preliminary examination, as well as
2. open needs (perceived deficits) about the question of possible support when driving, and
3. factors influencing the specific need, i.e., the willingness to spend money to remedy the deficits:

a) knowledge about driver assistance systems and assessment of one's own knowledge,
b) experience with driver assistance systems and assessment of one's own experiences,
c) evaluation of driver assistance systems, i.e., assessment of their advantages and disadvantages,
d) expectations towards driver assistance systems, and
e) customer wishes.

1. Indications of latent needs

In the preliminary examination, latent needs (deficits not yet perceived) were asked through an open question regarding experiences with age-related physical limitations. The responses show that most older drivers do not perceive age-related physical limitations. Almost half report not having noticed any limitations while driving (see Fig. 7a). Although this is a significantly lower proportion (at α <0.001) than in the comparison group of younger drivers, where over 80% report not having experienced any limitations. Given the medically and psychologically based age-related deterioration of cognitive, visual, and motor skills, the limitations due to environmental influences, and the ability to perceive, however, it would have been expected that most older drivers have already noticed limitations while driving.

Most frequently, older drivers mention physical limitations (28%), fewer restriction of senses (seeing, hearing, 16%), limitations due to environmental influences such as darkness and rain (11%), and limitations of perception ability (seven percent). In the comparison group, the order is the same, but the proportion of younger drivers who have experienced these limitations is lower in each case (seven, six, three, and two percent, respectively). Significant differences between the study and comparison groups exist only for physical limitations (α = 0.001) and somewhat less strongly for limitations of senses (α = 0.05).

Experiences with age-related physical limitations in the three age groups (50 to 59 years, 60 to 69 years, and 70 years and older) decrease with increasing age, contrary to expectations. The differences between the age groups are highly significant in terms of limitations of senses (with α <0.001, see Fig. 7b). Since the subjective perception was asked about, the results can be explained by the fact that older drivers suppress deficits and may even have become accustomed to them with increasing age, which is consistent with other studies (Rudinger 2015). It is suspected that the proportion of older drivers with limitations is actually much higher than in the survey. No significant difference is noticeable between male and female respondents. Only in the case of limitations due to environmen-

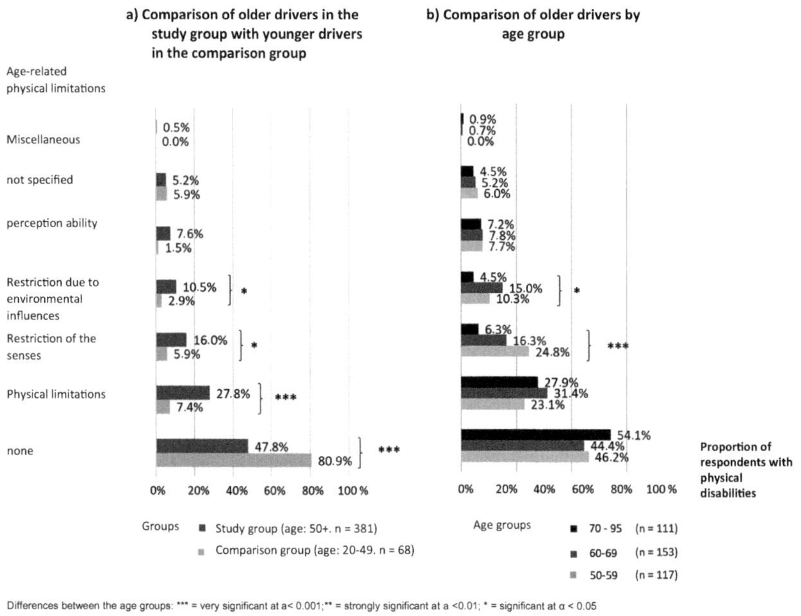

Fig. 7 Experiences with age-related physical limitations (preliminary examination, n = 449)

tal influences (darkness, rain), women seem to have more problems than men ($\alpha < 0.05$).

2. Indications of open needs

The question about open needs is denied by half of the older drivers (50 years and older) (Fig. 8a). Only 42% state that they do not need assistance while driving, not significantly more than in the comparison group (43%). Older drivers see potential support for themselves while driving mainly in terms of vision (14%), parking (13%), perception and driving (each 11%), braking only five percent. The differences to the comparison group of younger drivers are rather small and not significant. Older drivers see more support potential mainly in terms of vision and parking, younger drivers unexpectedly in terms of perception.

A closer look at older drivers by age groups (50 to 59 years, 60 to 69 years, 70 years and older) shows no significant differences except for parking, where the age group of 60- to 69-year-olds sees more need for support (see Fig. 8b).

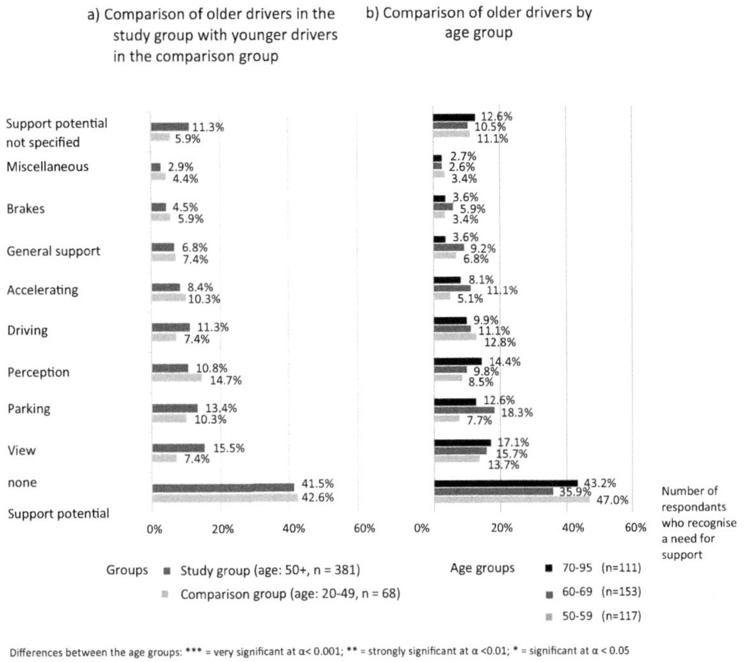

Fig. 8 Potential support while driving (preliminary examination, n = 449)

This age group also sees more need for support in braking and accelerating than the other two age groups and less often no need for support at all. Significant differences between male and female respondents are also not evident here. In summary, it can be stated that older drivers show a latent and open need for driver assistance systems, but to a lesser extent than would have been expected given physical and psychological limitations. Especially in terms of view, parking, and perception, a need for support is seen.

3. Indications of specific need for driver assistance systems

Indications of the specific need, i.e., the willingness of older drivers to spend money on driver assistance systems to remedy limitations, are provided by a) product knowledge (knowledge of driver assistance systems and assessment of one's own knowledge), b) experiences with driver assistance systems and assessment of one's own experience, c) evaluation of driver assistance systems, i.e., assessment of their advantages and disadvantages, d) expectations of driver assistance systems, and e) customer wishes.

a) Product knowledge
Driver assistance systems are known to over 90% of the 381 older drivers (50 years and older) and almost 95% of the 68 younger drivers. Given their potential for support, driver assistance systems that positively influence view should be sought after and therefore particularly known to older drivers. However, this is not the case. The most well-known among older drivers are parking aids (over 40%), cruise control (38%), distance and brake warning systems (38%) ahead of lane keeping assist (28%), navigation system (28%) and (emergency) brake assist (23%). ABS, ESP and ASR are also mentioned, although they are no longer considered driver assistance systems today. In the comparison group of younger drivers, the proportion of those who know the individual driver assistance systems is consistently larger (except for the navigation system, which is now standard in most vehicles). They are particularly familiar with the modern driver assistance systems (lane keeping assist, (emergency) brake assist, lane change and blind spot assist at $\alpha < 0.001$, 0.01 or 0.05, see Fig. 9a).

With increasing age, knowledge of driver assistance systems decreases (see Fig. 9b). This is significant (with $\alpha < 0.01$) only for cruise control, which 47% of 50 to 59-year-olds know and only 26% of respondents aged 70 and over. Women are less familiar with driver assistance systems, with the exception of parking aids, than men. For modern driver assistance systems (lane change and blind spot assist, distance and brake warning assist, and (emergency) brake assist), these differences are significant (at $\alpha < 0.001$ or $\alpha < 0.01$). In summary, it can be stated that knowledge of driver assistance systems among older drivers is rather low. With increasing age, product knowledge continues to decrease.

This result is also confirmed by the subjective assessment of the surveyed drivers regarding their knowledge of driver assistance systems, which was queried on a seven-point Likert scale from 1 "very low" to 7 "very high". With an average product knowledge of 3.005, the product knowledge of older drivers can be considered rather low. The product knowledge of younger drivers (comparison group) is not significantly higher (see Fig. 10). There are no significant differences between the individual age groups, even though the product knowledge of 60 to 69-year-olds is higher than that of the slightly younger and the even older drivers. Significant differences are only shown between men (3.543) and women (2.147).

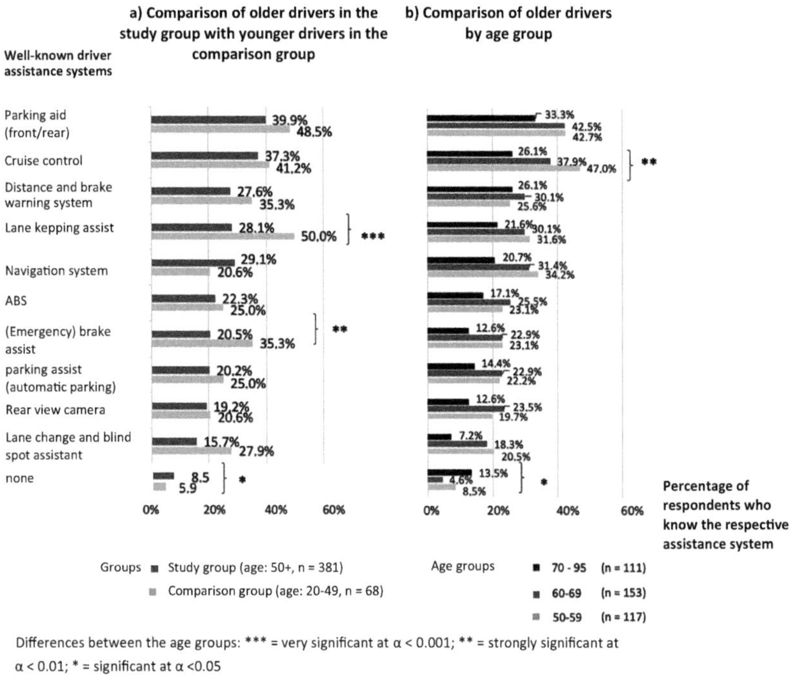

Fig. 9 The most well-known driver assistance systems (preliminary examination, n = 449)

b) Experiences with driver assistance systems

Although knowledge of driver assistance systems is still low, experiences with driver assistance systems are almost exclusively positive. They are considered useful (73%). Only three percent of respondents report having had a negative experience with a driver assistance system, e.g. because functions were unclear. 17% of the subjects have not yet had any experience with driver assistance systems. Here, the difference between the age groups is significant ($\alpha < 0.05$). Older test subjects more often have no experience with assistance systems (Fig. 11a and b).

This result confirms the assessment of the surveyed drivers regarding their experiences on a seven-point Likert scale from 1 "very low" to 7 "very high". They can be considered rather low in both the study group (3.368) and the comparison group (3.242) and do not differ significantly (Fig. 12). There are also no significant differences between the age groups, even though the experiences of the 60 to 69-year-olds are greater than those of the slightly younger and the even

Fig. 10 Assessment of own product knowledge (preliminary examination, n = 449)

Group	n	Mean values	Significance*
Control group	**68**	**3,353**	n.s. (Control group vs. Comparison group)
Comparison group	**381**	**3,005**	
Age group 50 - 59	117	2,872	n.s. (50 - 59 vs. 60 - 69); n.s.
Age group 60 - 69	153	3,196	n.s. (60 - 69 vs. 70 +)
Age group 70 +	111	2,883	
Men	293	3,543	sig. a <0.001
Women	156	2,147	

* Two-sample test for the difference between two arithmetic means

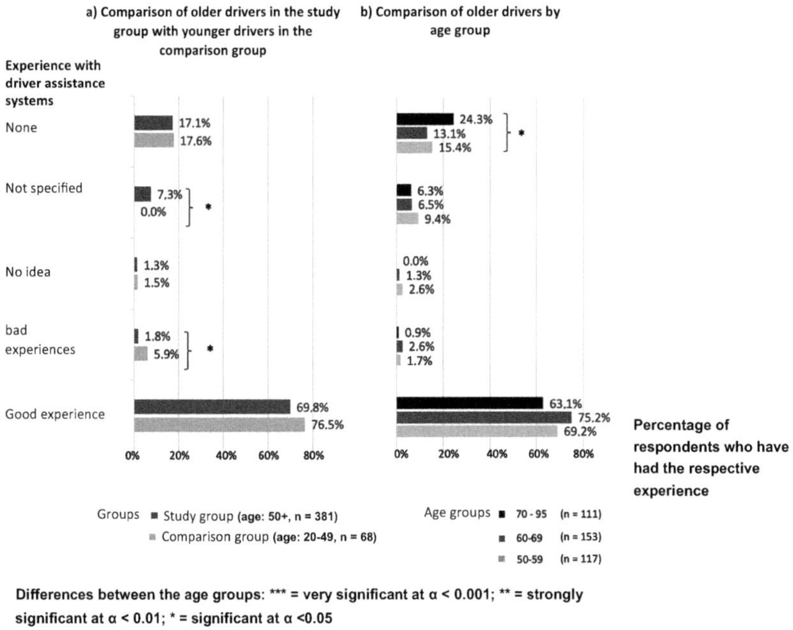

Fig. 11 Experiences with driver assistance systems (preliminary examination, n = 449)

Fig. 12 Assessment of own experience with driver assistance systems (preliminary examination, n = 449)

Group	n	Mean values	Significance*
Comparison group	**68**	**3,368**	n.s.
Study group	**381**	**3,242**	
Age group 50 - 59	117	2,966	n.s.
Age group 60 - 69	153	3,510	n.s.
Age group 70 +	111	3,164	n.s.
Men	293	3,613	sig. α <0,001
Women	156	2,603	

* Two-sample test for the difference between two arithmetic means

older drivers. Significant differences are only shown between men (3.613) and women (2.603).

Overall, the participants have had very positive experiences with driver assistance systems (e.g., Aral AG 2015; Oxley and Mitchell 1995; Stevens 2012; Son et al. 2015). Therefore, it is surprising that so few drivers are familiar with driver assistance systems.

c) Evaluation of driver assistance systems (assessment of their advantages and disadvantages).
Despite the rather limited experience with driver assistance systems, the participants are aware of the advantages of such systems, especially the increased safety in road traffic (57%). 56% see the advantage of support and relief as well as 35% a gain in comfort, only 11% a compensation for physical deficits. The younger drivers mention significantly more advantages compared to the study group, as can be seen from Fig. 13a. Especially the increased safety and accident prevention is highly rated by the comparison group (77%), significantly higher than the study group (53%). But also with "warning of dangers" and "gain in comfort" there is a highly significant difference between older drivers (6% and 33%) and younger ones (16% and 50%). There is also a significant difference in the "compensation of physical deficits" (10% and 19%).

The division by age groups shows that fewer advantages are seen with increasing age. The differences are only significant for "gain in comfort" and "compensation for physical deficits". The perception of the advantages differs between men and women only significantly for "compensation of physical deficits". Over-

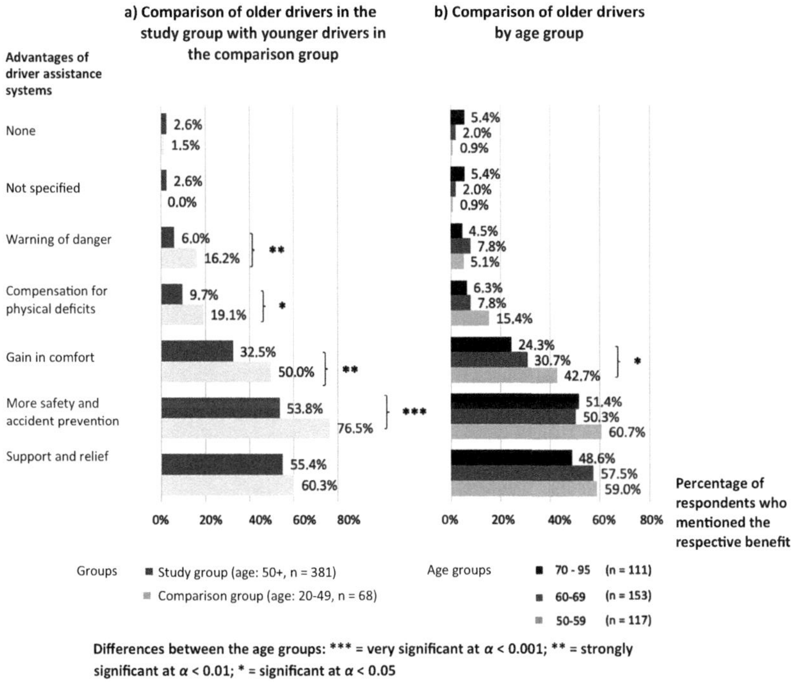

Fig. 13 Advantages of driver assistance systems (preliminary examination, n = 449)

all, clear advantages of driver assistance systems are seen, which is surprising given the limited knowledge and experience.

As disadvantages, the participants mention that they would have to rely too much on the technology (50%), also the risk of inattention and distraction (38%) as well as a loss of personal responsibility (22%). A detailed evaluation shows that younger drivers see significantly more disadvantages than older ones (see Fig. 14a). Above all, they too fear having to rely too much on the technology (62%) and losing personal responsibility (29%), but they also fear malfunctions (31%). The differences between younger and older drivers are highly significant when it comes to mentioning malfunctions. This also applies to costs and dependence on technology. Although both groups have had similar positive experiences, younger people have significantly more concerns than older ones.

In the individual age groups, different disadvantages of driver assistance systems are perceived differently strongly (Fig. 14b). The 50–59 year olds mainly

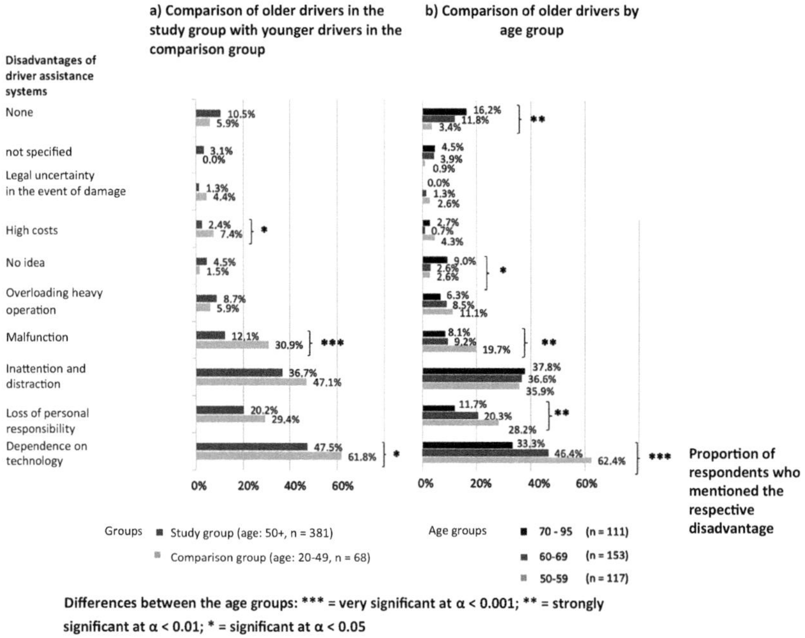

Fig. 14 Disadvantages of driver assistance systems (preliminary examination, n = 449)

mention the dependence on technology (very significant difference to the other age groups), the loss of personal responsibility (highly significant) and malfunctions (highly significant) as disadvantages. The 60–69 year olds and even older ones generally see fewer disadvantages. It is striking that all groups mention the risk of inattention and distraction. The older the participants are, the more likely they are to see no disadvantages (highly significant difference), which can be explained by limited knowledge and experience with driver assistance systems. There are no relevant significant differences between men and women.

d) Expectations of driver assistance systems

37% of the participants mainly expect greater safety from driver assistance systems. They should provide support (38%), increase comfort (24%) and be reliable (16%). Some participants have no expectations at all of driver assistance systems, five percent expect autonomous driving. In general, younger people expect more from them than older ones. The differences in expectations regarding the reliability of driver assistance systems are very significant. There are hardly any differences between men and women.

e) Customer wishes
The participants' wishes were also asked, in particular which driver assistance systems they would like to use and thus buy. By far the most older drivers want distance, brake and speed warning systems (28%), 24% a parking aid. The relatively low customer wishes are surprising, as e.g. many drivers have problems with parking (see Fig. 8a and b) and especially older drivers are familiar with parking aids. This also applies to the (emergency) brake assistant (18%), the cruise control (16%), the lane keeping assistant (14%) and the navigation system (14%). Despite the predominantly positive experiences with driver assistance systems, only relatively few participants want to use driver assistance systems. 17% of the participants do not want to use any driver assistance systems at all or cannot say anything about it. There are hardly any differences between younger and older ones.

Regarding customer wishes, there are significant differences between men and women. Men desire ABS, ESP, and ASR more strongly. Distance, brake, and speed warnings, parking aids, cruise control, lane keeping assistant, and lane change assistant, women, on the other hand, prefer parking aids. Distance, brake, and speed warnings, as well as parking aids, are the clearly preferred driver assistance systems that the respondents would use.

Further results of the survey on customer needs, customer satisfaction, and the need for safety are:

- only 30% of the test subjects, especially younger ones, consider driver assistance systems to be useful for motorway and overland journeys, even fewer for parking (21%), in traffic jams (20%), for controlling the distance (19%), in emergency braking situations (20%) or when overtaking (19%). Driver assistance systems are considered even less useful for city driving (13%).
- More than 60% of the subjects feel safer with driver assistance systems, 27% more relaxed and comfortable. 13% feel insecure. The age group of 50–59 year olds feels safer and more relaxed, but describes the use as very significantly more unusual and insecure than the other two age groups. Many of those over 70 cannot answer the question or do not comment on it.
- Women mention insecurity and the unusual use of driver assistance systems significantly more often and more strongly than men. Despite a positive attitude, they associate assistance systems with rather negative associations.

- Driver assistance systems primarily satisfy a need for safety, more than the need for comfort, for relinquishing control, and for warning of dangers. It is striking that a large number of subjects cannot assess their needs. The ranking is the same between the test and comparison group, but the need for safety and control is very significantly more important to younger drivers (57%) than to older drivers (37%). The older the subjects, the less they express needs.
- Reliability and low costs are important to the drivers surveyed, significantly more so to younger people than to older ones.

4 Changes in the Sample in the Main Examinations

The demographic characteristics of the test subjects of the three main examinations are now compared with the test subjects of the preliminary examination (Sect. 2).

1. Demographic characteristics of the sample in the first main examination
Of the 449 respondents in the preliminary examination (381 older and 68 younger drivers), 181 older (50 and older, average age 65 years) and 46 younger drivers (between 20 and 49 years, average age 28 years) participated in the first main examination with tests in the driving simulator. The demographic characteristics are similar, however, the proportion of those over 70 is lower in the first main investigation (see Fig. 15).

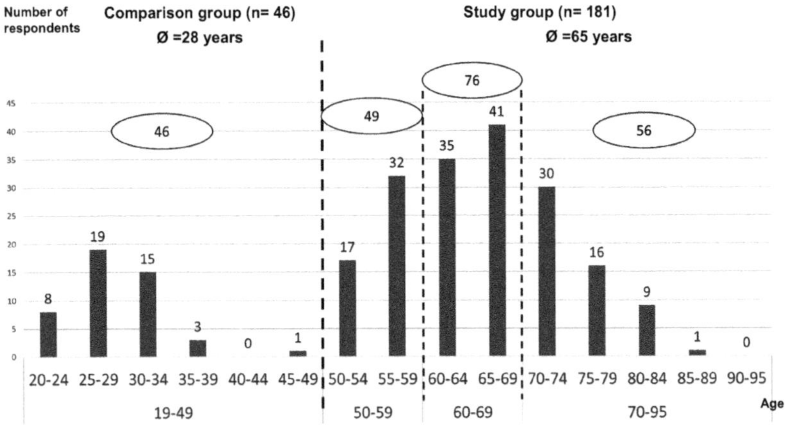

Fig. 15 Age distribution of the sample of the first main examination (n = 227)

The biggest differences to the preliminary examination are in the ratio of men to women, 80% of the older subjects in the driving simulator are male (in the preliminary examination 66%), 61% of the younger subjects (in the preliminary examination 59%).

Regarding educational qualifications, there are hardly any differences between the test subjects of the preliminary examination and the first main examination, likewise regarding net incomes. The average monthly net income is slightly higher in the second main examination, with about 3,500 EUR in the age group 50–54 years being the highest, in the preliminary examination the highest with about 3,000 EUR in the age group 60–60 years. The average driving performance of the subjects is between 21 and 50 km per day in both examinations.

2. Characteristics of the Sample for the Second Main Examination

Of the 449 (381 older, 68 younger) respondents from the preliminary examination, only 152 older and 31 younger drivers participated in the second examinaiton of improved acoustic driver assistance systems in the driving simulator. Fig. 16 shows the age distribution of the sample for the second main examinaiton: in the study group of older drivers, the average age is 66 years, in the comparison group with younger drivers, 29 years. In the study group, 23% are between 50 and 59 years old and 43% are between 60 and 69 years old, a third are 70 years and older, of which more than half are between 70 and 74. Eleven drivers are 80

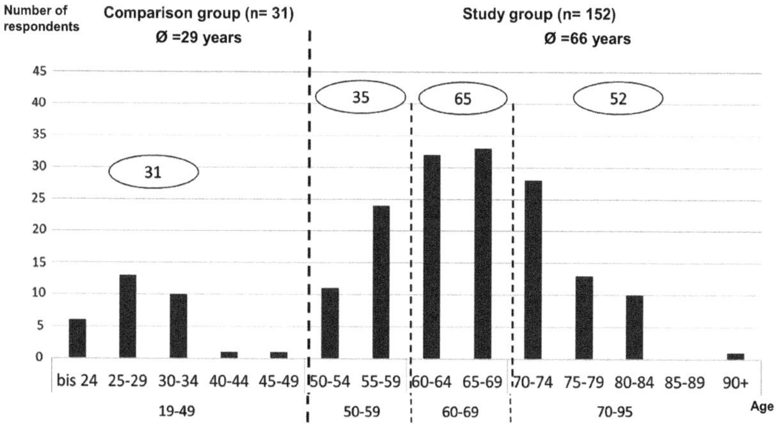

Fig. 16 Age distribution of the sample in the second main examination (n = 183)

years and older. The demographic characteristics of the two examinaitons (preliminary examinaiton and second main examination) are thus very similar.

Even higher than in the first main examination and thus much higher than in the preliminary examinaiton was the proportion of male test subjects in the second main examination with 82%, in the comparison group (younger drivers) 61% are still male. The distribution of educational qualifications is very similar. The average monthly net income of the test subjects of the second main examination is again somewhat higher than in the preliminary examination, highest with more than 3,600 EUR in the age group 60–64 years. The average driving performance remains unchanged between 21 and 50 km per day. The youngest drivers drive the least (about 22 km), the 35–50 year olds the most (about 60 km). With increasing age, the daily driving performance decreases.

3. Characteristics of the Sample for the Third Main Examinaiton

The sample of the third main examiniation with a drive on a test track with 46 test subjects (31 older, 15 younger drivers) is not comparable with the previous examinations, because due to the somewhat more distant test track some new test subjects had to be sought. The test subjects were interviewed after the drive on the test track in a vehicle with the five acoustic driver assistance systems improved for the second main examination. Figure 17 shows the age distribution of the sample of the third main examinaiton with an average age of 66 years in three

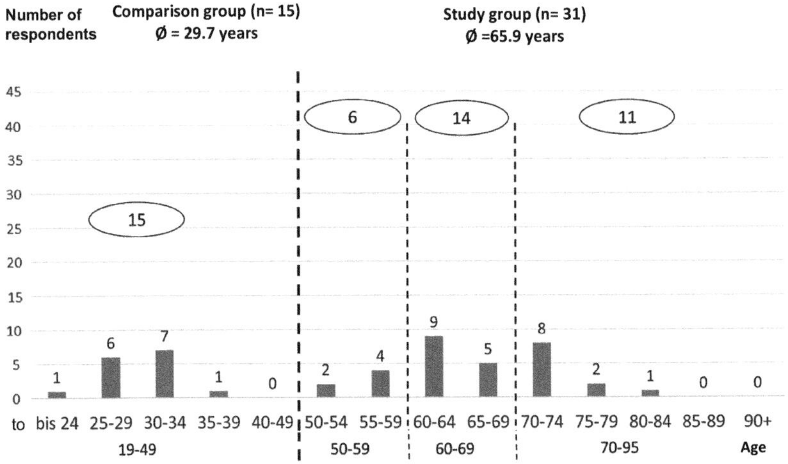

Fig. 17 Age distribution of the subjects on the test track (third main examination, n = 46)

age groups: six test subjects in the age group 50–59 years (20%), 14 test subjects in the age group 60–69 years (45%) and eleven test subjects in the age group 70 years and older (35%). The comparison group of younger drivers with an average age of 30 years has two subgroups: seven test subjects in the age group under 30 years and eight in the age group 30–39 years. Thus, the age distribution of the test drive is very similar to the examinations in the simulator.

Compared to the investigations in the simulator, the ratio of men to women on the test track is significantly more balanced (58 and 42% respectively). This distribution also corresponds to the distribution in the comparison group of younger drivers.

The test subjects on the test track, especially in the comparison group, have on average slightly higher educational qualifications than the test subjects in the examinations in the simulator. The net incomes of the test subjects of the study group (31 older drivers with an average income of about 2,240 EUR) also increase in this examination with increasing age. They are highest with 2,000 to 3,000 EUR in the age group 50–59 years. The net incomes in the comparison group of younger drivers are between 1000 and 2000 EUR. The average driving performance of the test subjects on the test track corresponds to the driving performance of the subjects of the previous investigations in the simulator, 15 to 50 km per day. The youngest test subjects drive the least, less than 20 km per day. The driving performance of the younger drivers, however, increases with increasing age to more than 50 km per day, the driving performance of the test subjects on the test track is about 40 km per day.

Thus, the demographic characteristics of the various examinaiton phases in the ALFASY project, in the simulator and on the test track, are similar.

5 Distinctive Features of Older Drivers

The various empirical examinations in the ALFASY project reveal some peculiarities of older drivers compared to younger drivers:

- the proportion of men interested in driver assistance systems is significantly higher than the proportion of women, and it increases with age. Female test subjects tend to express uncertainty about the driver assistance systems, which are often unfamiliar to them.
- The net income of older drivers is higher and increases with age (at least in the largest sample of the preliminary examination),
- the daily driving performance decreases with increasing age,

- older drivers express a latent and open need for driver assistance systems, albeit to a lesser extent than would have been expected given physical and psychological limitations. They seek assistance with vision, parking, and perception of the vehicle environment.
- The specific needs of older drivers are still under-researched:
  - their knowledge of driver assistance systems is not very high and decreases with increasing age, their experiences are (like those of all respondents) overall positive, older drivers see clear advantages of driver assistance systems, even if they have not yet had their own experiences.
  - They are reluctant to rely on technology, less so than younger drivers, and are reluctant to give up personal responsibility, which is not surprising given their limited experience with driver assistance systems.
  - All drivers rarely express wishes for driver assistance systems. However, it is always difficult to express wishes for technical devices and systems when they are not known. This is especially true for older drivers, who know little about offers and utility value. They do not know their needs or cannot articulate them, the older the less.
  - For older drivers, low costs are less important than for younger ones.

The study reveals a large untapped market potential. Therefore, powerful assistance systems for older drivers are important (Chapter "Driver Assistance Systems—An Overview" to "Evaluation of the Performance of Acoustic Assistance Systems for Older Drivers"), which improve driving behavior (Chapter. "Driving Behavior of Older People in the Driving Simulator"), but also driving performance and stress behavior (Chapter "Neuropsychological and -physiological Correlates of the Driving Behavior of Older Drivers within Simulated Environments") and for which there is customer acceptance and willingness to pay (Chap. "Market Potentials of Older Drivers - Willingness to Pay and Acceptance of Age-appropriate Driver Assistance Systems"). For this, they first need to become better known.

References

ARAL AG (2015). Aral Studie – Trends beim Autokauf 2015. Marktforschungsbericht der ARAL AG. Bochum.

Brooks, J. O., Goodenough, R. R., Crisler, M. C., Klein, N. D., Alley, R. L., Koon, B. L., Logan Jr., W. C., Ogle, J. H., Tyrrell, R. A., & Wills, R. F. (2010). Simulator sickness during driving simulation studies. *Accident Analysis and Prevention*, *42*(3), 788–796.

Homburg, C., & Krohmer, H. (2006). *Marketingmanagement: Strategie – Instrumente – Umsetzung – Unternehmensführung*. Wiesbaden: Gabler.

Homburg, C. (2014). *Grundlagen des Marketing-Managements: Einführung in Strategie, Instrumente, Umsetzung und Unternehmensführung*. 4th edn. Wiesbaden: SpringerGabler.

Kohlbacher, F., & Herstatt, C. (eds.). (2011). *The silver market phenomenon. Marketing and innovation in the aging society*. Berlin: Springer.

Oxley, P. R., & Mitchell, C. G. B. (1995). *Final report on elderly and disabled drivers´ information telematics (Project EDDIT)*. Brussels, Belgium: Commission of the European Communities DG XIII, R & D Programme Telematics Systems in the Area of Transport (DRIVE II).

Roenker, D. L., Cissell, G. M., Ball, K. K., Wadley, V. G., & Edwards, J. D. (2003). Speed-of-processing and driving simulator training result in improved driving performance. *Human Factors, 45*(2), 218–233.

Rudinger, G. (2015). Zielgruppe Seniorinnen und Senioren. In C. Klimmt, M. Maurer, H. Holte, E. Baumann (eds.), *Verkehrssicherheitskommunikation: Beiträge der empirischen Forschung zur strategischen Unfallprävention (S. 53–79)*. Wiesbaden: Springer.

Schweig, S., Liebherr, M., Schramm, D., Brand, M., & Maas, N. (2018). The impact of psychological and demographic parameters on simulator sickness. *Proceedings of 8th International Conference on Simulation and Modeling Methodologies, Technologies and Applications (SIMULTECH 2018)*, 91–97.

Son, J., Park, M., & Park, B. B. (2015). The effect of age, gender and roadway environment on the acceptance and effectiveness of advanced driver assistance systems. *Transportation Research Part F: Traffic Psychology and Behaviour, 31*, 12–24.

Stevens, S. (2012). The relationship between driver acceptance and system effectiveness in car-based collision warning systems: Evidence of an overreliance effect in older drivers? *SAE International Journal of Passenger Cars – Electronic and Electrical Systems, 5*(1), 114–124.

Winner, H., & Schopper, M. (2015). Adaptive cruise control. In H. Winner, S. Hakuli, G. Wolf (eds.), *Handbuch Fahrerassistenzsysteme – Grundlagen, Komponenten und Systeme für aktive Sicherheit und Komfort*. Wiesbaden: Vieweg+Teubner, 851–891.

Driver Assistance Systems—An Overview

Dieter Schramm and Stephan Schweig

Contents

The term driver assistance system (DAS) describes a system whose task is to assist the driver in various areas of vehicle control, that is, in the interaction between driver, vehicle, and environment (Fig. 1), characetrized in particular by the traffic situation.

Prof. Dr.-Ing. Dieter Schramm, Dipl.-Ing. Stephan Schweig, all University of Duisburg-Essen.

D. Schramm · S. Schweig (✉)
Lehrstuhl für Mechatronik, Universität Duisburg, Essen, Germany
e-mail: stephan.schweig@uni-due.de

D. Schramm
e-mail: dieter.schramm@uni-due.de

H. Proff et al. (eds.), *Age-appropriate Driver Assistance Systems*,
https://doi.org/10.1007/978-3-658-44838-7_3

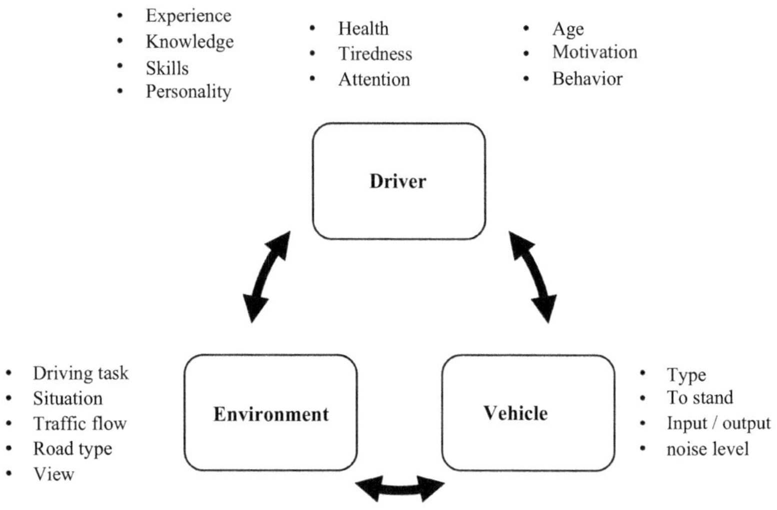

Fig. 1 Interaction of driver, vehicle, and environment, according to. *Source:* König, 2015

Today, the term is associated with both safety-enhancing and comfort-enhancing systems. Directly safety-enhancing systems include vehicle dynamics control systems that serve to stabilize the vehicle. Numerous driver assistance systems, especially in the area of vehicle dynamics control systems, have been standard in vehicles for many years, such as the anti-lock braking system (ABS) or the extension Electronic Stability Control (ESP). Depending on the interpretation, systems such as power steering are also considered assistance systems.

A definition of driver assistance systems is given by (Reif, 2010):

> "Driver assistance systems support the driver in his primary driving task. They inform and warn him, increase his comfort and safety by actively supporting him in his vehicle control and vehicle stabilization. If necessary, they reduce his workload."

This chapter provides a brief introduction to the topic of driver assistance systems, describes the assistance systems used in the ALFASY project, and briefly explains their functionality.

1 Motivation

In addition to increasing comfort, enhancing traffic safety is also a driving force in the development of DAS. Vehicle dynamics control systems, which do not aim at increasing comfort, usually serve the direct and indirect stabilization of the vehicle (by supporting the driver) to prevent accidents or reduce their severity.

The direct and indirect support potentials of driver assistance systems in the field of safety, which go beyond the functionalities of vehicle dynamics control systems such as ABS and ESP, can be easily explained using accident statistics (Fig. 2). The graph shows the trend of the number of fatalities in road traffic accidents in Germany since the early 1950s. Legislative measures that influenced the accident numbers are listed above the curve. The driver assistance systems and their predecessors introduced over the years are shown below the curve. The reductions in fatalities, which naturally occur with a time delay, can be seen.

According to (Brown, 2002), 95% of all road traffic accidents are caused by human error. Here, DAS help by autonomously performing selected driving tasks and relieving the driver, especially in routine tasks. This reduces the number of tasks the driver has to perform, which reduces the potential for human error. In addition, the driver becomes less fatigued, which also reduces the risk of human error.

The proven positive effect of some of these systems, which have been used in production vehicles for years, has already been reflected in legislation. For

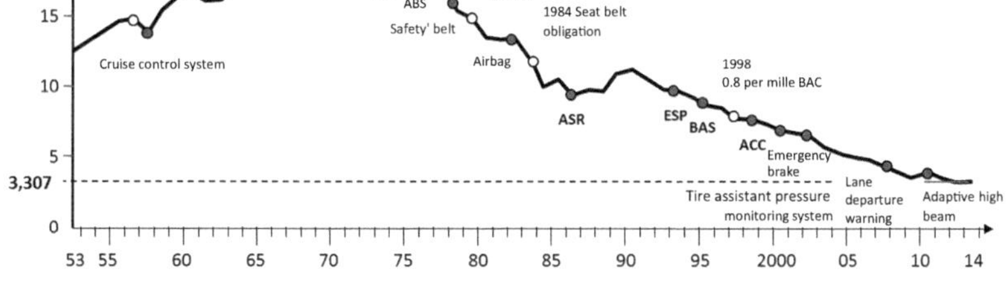

Fig. 2 Motorized vehicles and accidents since 1970

example, in Germany, vehicles with a total mass of more than 3.5 tons must be equipped with an anti-lock braking system (ABS), (Federal Ministry of Transport, 2011). However, current legislation is still significantly behind technical development, which varies from region to region. As part of the harmonization of rules by the EU, new DAS are continuously declared standard for new cars.

The development of new assistance systems, especially through assistance systems based on vehicle dynamics control, in recent years has drastically reduced the number of accidents, especially those with fatalities, as can be seen from (Fig. 2). At the same time, the degree of automation of driver assistance systems has dramatically increased with the rapid technical development in the central areas of sensor technology, actuator technology, and algorithms (Fig. 3). Figure 3 also shows the influence of the respective sensor systems used and the trend towards connectivity in road traffic. The development of algorithms is also being vigorously pursued. Here, methods of artificial intelligence, such as machine learning, are increasingly being used on a large scale, see e.g. (Rehder, Koenig, Goehl, Louis, & Schramm, 2019).

This, however, based on a longer time horizon, also suggests a development towards fully automated driving.

A systematic classification of the degree of automation of driver assistance systems was established by the SAE in the standard SAE J3016 (SAE, 2014) as

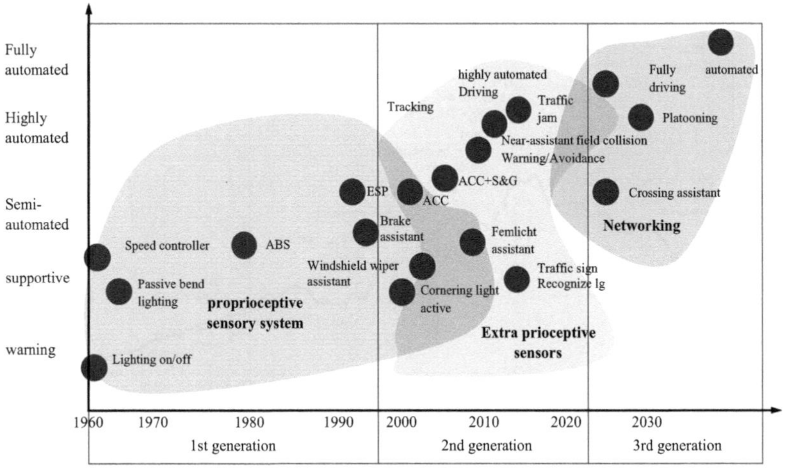

Fig. 3 Degree of automation of driver assistance systems in temporal development. *Source:* Schramm, Hesse, Unterreiner, & Maas, 2017

shown in Fig. 4. The assistance systems currently available in production vehicles are at most assignable to level 2 of the classification. Level 3 systems are in development, but currently do not yet have production readiness and no road approval. A complication here is that the driver must regain control of the vehicle in the event of a system drop, e.g., in traffic situations that the system can no longer handle (Maas, 2017), (Maas & Schramm, 2016). It must be ensured that he is able to do so and it must be clarified how the time until takeover is bridged.

The penetration of newly registered vehicle populations with driver assistance systems of levels 1 and 2 is quite remarkable. An overview of the installation rates of the most commonly used DAS in the years 2015 and 2016 is shown in Fig. 5. This development is increasingly driven by legislation. For example, a new law was initiated at the European level that mandates a number of assistance systems as standard for new vehicles from 2022. In the passenger car sector, these include an intelligent speed assistant, a driver fatigue warning system, a reversing assistant, an emergency brake assistant to warn following vehicles, and the preparation of an alcohol breath test to prevent vehicles with intoxicated drivers from driving off, as well as an event-related data recording device (black box) (European_Parliament, 2019).

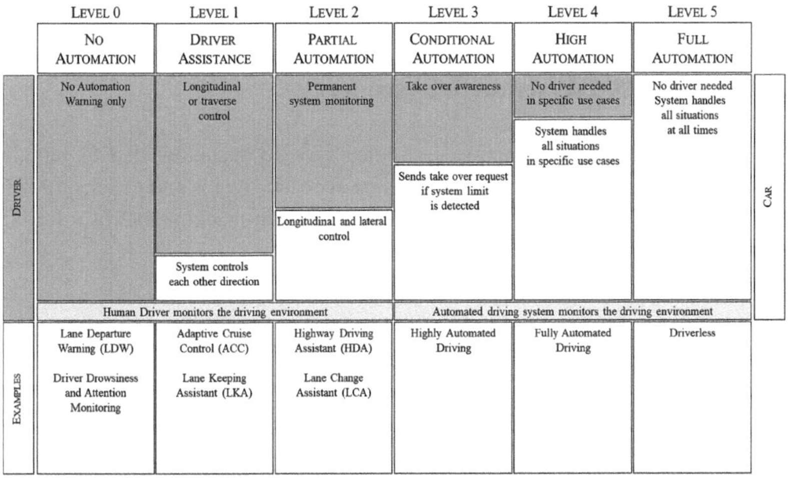

Fig. 4 Levels of automated driving based on SAE J3016 (SAE, 2014) *Source* Schramm, Hesse, Maas, & Unterreiner, 2020

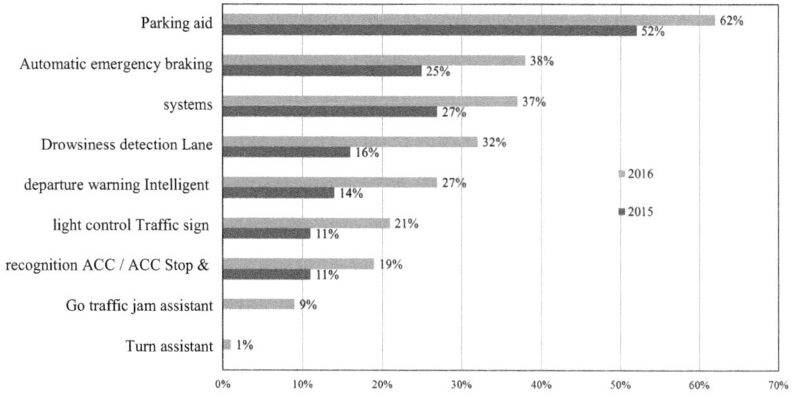

Fig. 5 Installation rates of FAS for new cars in Germany for the years 2015 and 2016. (Data source: Bosch, 2018)

This article looks at a selection of systems that can particularly support drivers with age-related restrictions in safely and comfortably operating their vehicles. In particular, it should be investigated which human-machine interfaces are suitable for fully exploiting the potential of the assistance systems.

2 Selected Assistance Systems

In the ALFASY project, suitable assistance systems were selected to initially make these perceptible to older people using acoustic signals and to scientifically investigate the benefit of these systems and the human-machine interface for this population group.

2.1 Cross Traffic Alert when Reversing

The cross traffic alert monitors a partial area of the space behind the vehicle when reversing.

As shown in Fig. 6, a larger space behind and to the side of the vehicle is monitored. The core area of monitoring is the traffic to the right and left of the

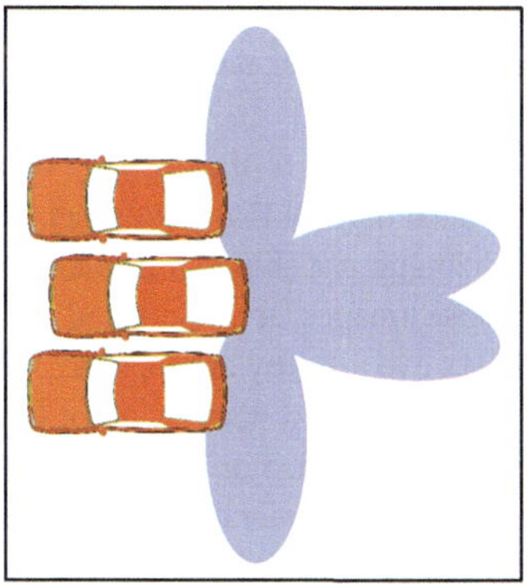

Fig. 6 Cross traffic alert when reversing

vehicle. That is, the system warns the driver visually and/or acoustically of crossing traffic behind the vehicle's rear. Range and coverage vary according to sensor selection and manufacturer. A space monitoring of up to 50 m can be achieved, for example, by two mid-range radar sensors, which are mounted on the left and right at the rear of the vehicle (Bosch, 2020).

2.2 Parking Assistance

The parking assistance system is one of the most well-known assistance systems. The system is based on environmental sensors such as ultrasound, radar, or cameras to detect the distances of objects around the vehicle. Estimating the vehicle geometry in the front and rear areas is the biggest challenge when parking. Aerodynamic and design-related shapes can limit visibility. This applies in particular to the shape and size of pillars and window surfaces. Parking assistance is also available in various forms in vehicles.

The most well-known and simplest principle is the ultrasound-based parking assistance system. This is a purely acoustic and informative system. The ultrasound sensors installed at the front and rear and sometimes at the side of the vehicle determine the distances to surrounding objects. The determined distance is usually communicated acoustically by an interval tone. The distance to the objects determines the gap between the tones. To distinguish the direction, the frequency of the interval tone can be changed in some systems. In addition, in vehicles with multiple speakers or other acoustic signal generators, a variable location for the sound source is possible, which can help with orientation in space.

In addition to the acoustic display, better-equipped modern systems also have a visual display that represents the distances with simple color markings around a schematic car (Fig. 7). It is also possible to combine this with multiple cameras, whose images are calculated and combined into an image from the (apparent) bird's eye view by appropriate control unit software.

Advanced systems can not only detect distances but also calculate action recommendations. For example, a system can display guide lines in the image of a rear-view camera and thus assist the driver in steering maneuvers. Other systems measure the parking space and suggest a sequence of steering maneuvers for parking.

Semi-automated and fully automated parking systems have also been available for several years. In semi-automated systems, the control unit plans the trajectory for parking and adjusts the steering wheel. The driver is still actively involved in the parking process by operating the drive and brake pedals. In fully automatic parking, this function is also taken over by a control unit of the vehicle. Some systems even allow a "remote operation", where the driver can leave the vehicle and observe the parking process from the outside. However, according to current legal rulings, the driver's supervision is mandatory for all fully automatic systems. A conceivable monitoring option for such a system is, for example, a "dead man's switch" that must be operated throughout the entire parking process.

Fig. 7 Visual representation of a parking assistance system

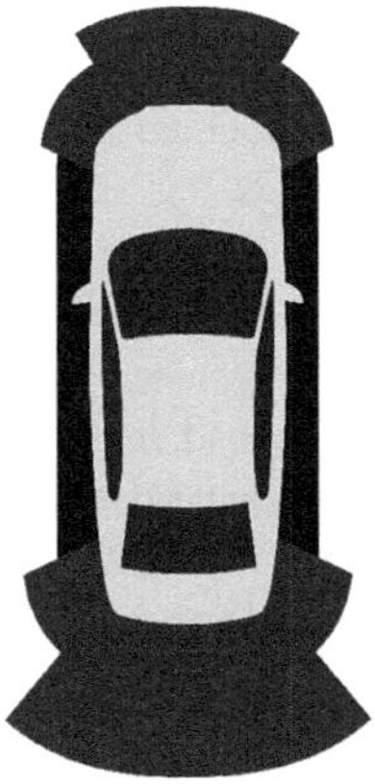

2.3 Blind Spot Assistant

Before changing lanes, the driver is obliged to check the rear and side areas of the vehicle. This is done by monitoring the mirrors and a shoulder check. If mirror settings are incorrect or a control measure is omitted, road users in the blind spot can be overlooked. A lane change in such cases leads to a collision. Another source of error is underestimating the speed of approaching vehicles from behind on country roads and highways (Winner, Hakuli, & Wolf, 2012).

To detect these dangers, ISO standard 17387 (ISO, 2008) provides for the monitoring of two zones. As Fig. 8 shows, both the direct areas next to the vehicle and a larger zone behind the vehicle must be monitored. Radar, Lidar, and cameras are also suitable for monitoring these areas. In current production vehicles, for example, monitoring is carried out by medium-range radar sensors, which are mounted on the side at the rear or in the exterior mirrors (Bosch, 2020).

The sensors detect objects directly next to the vehicle as well as vehicles at a greater distance. In the rear area, not only the occurrence of an object is detected, but also its relative speed. If objects are detected next to the vehicle, the driver is warned visually, e.g., by a lamp on the side mirror. This also applies if vehicles are approaching with an increased relative speed. Such turn assistants will become mandatory in the EU, initially for trucks and buses from 2022 (European_Commission, 2019).

2.4 Lane Keeping Assist

In addition to longitudinal control systems, such as ACC, which have been in series production for about 20 years, systems for lateral control in the lane guidance level are now also being used. One of these systems is the lane keeping

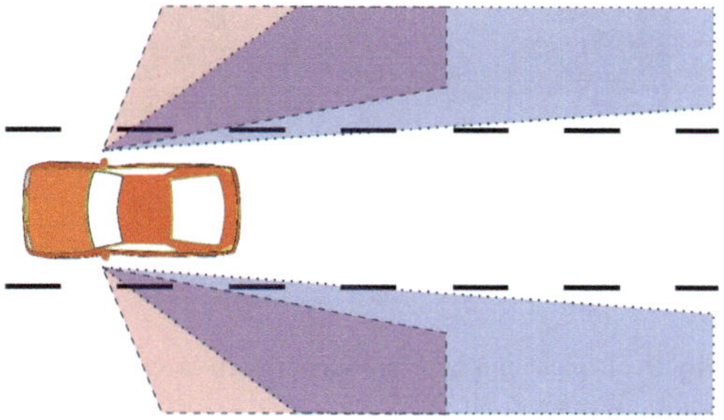

Fig. 8 Monitoring in two zones according to ISO 17387

assist, which is intended to prevent unintentional crossing of lane markings and edges. Basically, these systems can be divided into active and passive variants (Schramm et al., 2017):

- Systems whose task is to warn the driver when crossing the lane marking are referred to as passive lane keeping assistants. The warning before crossing the lane marking can be acoustic, visual or even haptic. Such systems are also referred to as lane departure warning systems.
- In addition to the warning function of passive systems, active systems can support the driver by an additionally set steering torque to adjust the vehicle alignment in relation to its own lane. These systems are often referred to as active lane keeping assistants.

Both variants of the lane keeping assistant are based on the same measurement and evaluation of their own position relative to the chosen lane. To maintain these relative distances and angles, a camera system measures the position of the lane markings. This requires a high technical level of the hardware used, as, unlike many applications of industrial image processing in production, the environmental conditions in road traffic are not controllable.

Lane detection systems receive their data from an environment-sensing sensor such as a camera, which is mounted inside the driver's cabin, for example behind the rear-view mirror, and observes the road. The data obtained are usually processed by various algorithms of image processing and/or methods of machine learning to detect lane markings in the space in front of the vehicle, as schematically shown in Fig. 9. Completely error-free image recognition under all weather and light conditions is currently impossible due to the complexity of the images

Fig. 9 Typical application scenario of a camera-based lane keeping assistance system. *Source* Winner et al., 2012

and situations that occur. Since the road surface is also not flawless, additional disturbances such as missing or faulty lane markings must be taken into account, as well as construction site markings.

Depending on the equipment of a lane departure warning system, it can signal a lane departure from the with different signals. This can be done visually by displays on the vehicle's instrument panel, acoustically by various signal tones or haptically by vibrations on the steering wheel or seat. In addition to the pure warning, many lane departure warning systems can now also perform short steering maneuvers and thus steer the vehicle back onto the lane independently. Such steering maneuvers are not necessarily carried out by a steering intervention, but it is also possible to initiate a twist back into the road into the vehicle by a torque resulting from braking the wheels on one side around the axis perpendicular to the road surface.

The highest level of development currently are so-called lane keeping assistants, possibly combined with lane change assistants. With these, the vehicle follows its lane independently or changes it, if the driver indicates this wish, for example by activating the direction indicator, through appropriate steering interventions in a wide range without driver interventions. According to current legal interpretation, it must be monitored at predetermined short intervals that the driver can take over the vehicle at any time and without delay. This is typically done by the driver touching the steering wheel regularly. An exception applies to so-called traffic jam assistants below a certain speed.

2.5 Tire Pressure Monitoring System

Since November 2014, all new vehicles within the European Union with a permissible total weight of up to 3.5 t must be equipped with Tire Pressure Monitoring Systems (TPMS) (EU Regulation No. 661/2009). A distinction is made between direct and indirect tire pressure monitoring systems. These differ in the way the inflation pressure is monitored.

- Direct tire pressure monitoring systems use a tire pressure sensor in each wheel to measure the inflation pressures and fill air temperatures (Fig. 10). This information is transmitted along with the individual sensor identification by radio signal (Europe 434 MHz band) to the control unit in the vehicle (Hoppe, Kessler, Müller, & Wagner, 2013). This allows a tire-specific pressure display in the instrument cluster. Direct measuring systems detect both slow diffusion losses and rapid pressure losses. Tire pressure sensors achieve a

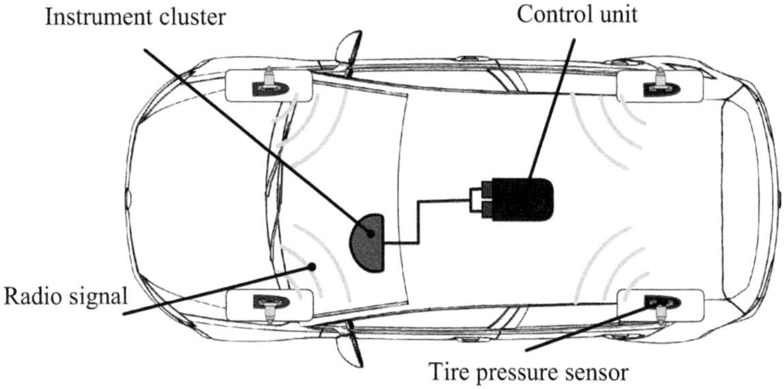

Fig. 10 Schematic of a direct tire pressure monitoring system. *Source* based on Hoppe et al., 2013

measurement accuracy of approx. 0.1 bar. The sensors typically deliver a signal every 30 to 60 s. A control unit processes the information and prepares it for display in the instrument cluster. The state of the art is the integration of the sensor into the rim. The unit consisting of the sensor housing and the valve screwed into it is supplemented by a circuit board with an integrated sensor and other electronic components, including a power supply. A new innovative concept for tire pressure sensor mounting is the attachment of the sensor to the inside of the tire tread.

- In contrast to direct tire pressure monitoring systems, indirect tire pressure monitoring systems do not measure the pressure in the tires, but primarily use the existing signals from the ABS wheel speed sensors (Fig. 11). These systems take advantage of the fact that the rolling circumference of the wheel and thus the effective rolling radius depends on the tire pressure. If the tire pressure drops, the rolling radius decreases. This in turn increases the speed of the wheel in relation to the other wheels. By evaluating the differential speeds of all wheels, the pressure loss of individual wheels can be detected. Advantages of indirect systems are lower costs, as no additional components need to be implemented. The disadvantage is that neither tire pressures nor the temperature of the fill air can be measured. A correct warning depends on the correct calibration of the system. During high-dynamic driving with lateral accelerations greater than $2\,m/s^2$ the system is inactive.

The display of the data determined by the system is done in a similar form as will be described later for the driving simulator used in the section.

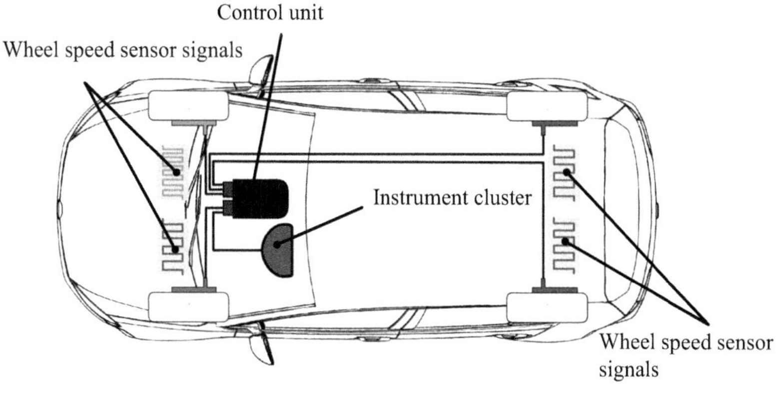

Fig. 11 Schematic of an indirect tire pressure monitoring system. *Source* based on Hoppe et al., 2013

References

Bosch. (2018). Anteil der in Neuwagen verbauten Fahrerassistenzsysteme in Deutschland in den Jahren von 2015 bis 2016. Retrieved from https://de.statista.com/statistik/daten/studie/510266/umfrage/anteil-der-in-neuwagen-verbauten-fahrerassistenzsysteme-in-deutschland/.

Bosch. (2020). Querverkehrswarnung. Retrieved from https://www.bosch-mobility-solutions.de/de/produkte-und-services/pkw-und-leichte-nutzfahrzeuge/fahrerassistenzsysteme/querverkehrswarnung/.

Brown, I. D. (2002). *A review of the "looked but failed to see" accident causation factor.* Paper presented at the Behavioural research in road safety: Eleventh Seminar, London.

Bundesministerium für Verkehr, B. u. S. (2011). Verordnung über die Zulassung von Fahrzeugen zum Straßenverkehr (Fahrzeug-Zulassungsverordnung – FZV). In.

Europäische_Kommission. (2019). Abbiegeassistenten für Lkw und Busse werden ab 2022 Pflicht. Retrieved from https://ec.europa.eu/germany/news/20190326-abbiegeassistenten_de.

European_Parliament. (2019). General safety of vehicles and protection of vulnerable road users. Retrieved from http://www.europarl.europa.eu/thinktank/de/document.html?reference=EPRS_BRI%282018%29625192.

EU-Verordnung (2019). *Verordnung (EG) Nr. 661/2009 des europäischen Parlamentsund des Rates vom 13. Juli 2009 über die Typgenehmigung von Kraftfahrzeugen, Kraftfahrzeuganhängern und von Systemen, Bauteilen und selbstständigen technischen Einheiten für diese Fahrzeuge hinsichtlich ihrer allgemeinen Sicherheit.* Received from https://eur-lex.europa.eu/LexUriServ/LexUriServ.do?uri=OJ:L:2009:200:0001:0024:DE:PDF.

Hoppe, U., Kessler, R., Müller, B., & Wagner, M. (2013). *Reifendruckkontrollsysteme*: Süddeutscher Verlag onpact GmbH.
ISO. (2008). ISO 17387: Intelligent transport systems – Lane change decision aid systems (LCDAS) – Performance requirements and test procedures. In.
König, W. (2015). Nutzergerechte Entwicklung der Mensch-Maschine-Interaktion von Fahrerassistenzsystemen. In *Handbuch Fahrerassistenzsysteme* (pp. 621–632): Springer.
Maas, N. (2017). *Konzeptionierung, Auslegung und Umsetzung von Assistenzfunktionen für die Übergabe der Fahraufgabe aus hochautomatisiertem Fahrbetrieb.* (Doktor Dissertation). Duisburg-Essen, Duisburg.
Maas, N., Hesse, B., Koppers, M., & Schramm, D. (2014, 10–12.09.2014). *Simulator setup according to use case scenarios-A human-oriented method for virtual development.* Paper presented at the Mechatronic and Embedded Systems and Applications (MESA), 2014 IEEE/ASME 10th International Conference on, Senigallia, Italien.
Maas, N., & Schramm, D. (2016, 01.03.2016-02.03.2016). *Lösungsansätze zur Problematik der Übergabe der Fahraufgabe an den Fahrer aus automatisiertem Fahrbetrieb.* Paper presented at the AmE 2016–Automotive meets Electronics, Dortmund, Deutschland.
Rehder, T., Koenig, A., Goehl, M., Louis, L., & Schramm, D. (2019). Lane Change Intention Awareness for Assisted and Automated Driving on Highways. *IEEE Transactions on Intelligent Vehicles, 4*(2), 265–276. https://doi.org/10.1109/tiv.2019.2904386.
Reif, K. (2010). *Fahrstabilisierungssysteme und Fahrerassistenzsysteme*: Springer.
SAE. (2014). SAE J3016 Taxonomy and Definitions for Terms Related to On-Road Motor Vehicle Automated Driving Systems. In. Berlin: Beuth.
Schramm, D., Hesse, B., Maas, N., & Unterreiner, M. (2020). *Vehicle Technology – Technical Foundations of Current and Future Motor Vehicles*. Berlin/Boston: De Gruyter Oldenbourg.
Schramm, D., Hesse, B., Unterreiner, M., & Maas, N. (2017). *Fahrzeugtechnik – Technische Grundlagen aktueller und zukünftiger Kraftfahrzeuge*. Duisburg: De Gruyter Oldenburg.
Winner, H., Hakuli, S., & Wolf, G. (2012). *Handbuch Fahrerassistenzsysteme: Grundlagen, Komponenten und Systeme für aktive Sicherheit und Komfort* (2., korrigierte Auflage ed.). Wiesbaden: Vieweg+Teubner Verlag/ Springer Fachmedien Wiesbaden GmbH Wiesbaden.

Driver Assistance Systems in the Context of Age-Appropriate HMI Design

Martin Brockmann, Milan Schreiber, Sophia Wingen
and Pia Immoor

Contents

The increasingly aging world population (DESA, UN, 2019; Haustein, Mischke, Schönfeeld, & Willand, 2016) and the associated increase in older road users (Haustein, Mischke, Schönfeeld, & Willand, 2016) already poses new challenges to the automotive industry now and even more so in the future. Studies

M. Brockmann (✉) · M. Schreiber · S. Wingen
Allround Team GmbH, Köln, Germany

P. Immoor
Allround Team GmbH, Erftstadt, Germany
e-mail: Pia.Immoor@allround-team.com

H. Proff et al. (eds.), *Age-appropriate Driver Assistance Systems*,
https://doi.org/10.1007/978-3-658-44838-7_4

show that the decline in sensory, physical, and cognitive abilities in old age brings many peculiarities with regard to participation in road traffic (Davidse, 2006) and involvement in traffic accidents (Young, Koppel, & Charlton, 2017; Fornells, Parera, Ferrer, & Fiorentino, 2017). To support this group, the use of driver assistance systems, the so-called Advanced Driver Assistance Systems (ADAS; Davidse, 2006), is recommended. However, the proportion of older people who actually own and use ADAS technologies is very low (Trübswetter 2015). A central reason for this could be problems with the usability and comprehensibility of the user interface. This user interface is referred to as the Human-Machine Interface (HMI). The HMI is the interface between the technology and the user and is therefore a decisive factor for the experience and frequency of use of assistance systems (Harvey, Stanton, Pickering, McDonald, & Zheng, 2011). Within the scope of this contribution, it should be worked out to what extent already available assistance systems are of benefit to older people and what problems older people have due to their age-related deficits in driving and interacting with user interfaces. Based on this analysis, ADAS and HMIs can then be designed that address the problems of this group of people (Kupschick, Bürgelen, Jürgensohn, & Protzak, 2019). In addition, technical advancements, such as improved and more reliable sensor technology and the increasing automation of vehicles, are expected to offer even greater support potential for assistance systems for older people in the future.

1 ADAS as Potential Aid

The group of older road users is particularly at risk of being seriously injured or killed in traffic accidents (Langford & Koppel, 2006). Especially in complex situations, such as intersections, it is evident that older drivers have an increased likelihood of being involved in accidents (Fornells, Parera, Ferrer, & Fiorentino, 2017). Although older people are involved in fewer accidents in absolute numbers than younger ones, they drive less overall with age and are thus more at risk per kilometer driven of injuring themselves and others (Trübswetter, 2015). Driver assistance systems offer a promising solution to increase driving safety (Davidse, 2006). According to the German Road Safety Council (DVR; 2016), driver assistance systems are defined as follows: "Driver assistance systems are systems that are suitable for supporting the driver in his driving task in terms of perception, driving planning and operation […]" (German Road Safety Council e. V., 2006). The lane departure warning, collision warning system, blind spot assistant, parking assistants, navigation systems, adaptive cruise control, automatic crash notification, and adaptive headlights were identified as helpful for older drivers within a literature review by Eby et al. (2016). In addition, older people saw a higher benefit in an intersection assistant than younger ones and stated that the system offers them the possibility to compensate for age-related deficits (Ziefle, Pappachan, Jakobs, & Wallentowitz, 2008). Assistance systems already available on the market thus prove to be helpful for older people. This was also shown in a study in which older participants, compared to younger participants, rated supportive systems in their own vehicles better and showed a higher willingness to buy them again (Crump, et al., 2016). In general, assistance systems can particularly support older drivers as they compensate for age-related weaknesses and thus improve driving safety (Ziefle, Pappachan, Jakobs, & Wallentowitz, 2008). The systems support the drivers, among other things, in reacting faster to upcoming obstacles, avoiding collisions, or keeping the lane safer even when distracted. They can also reduce mental demands and stress (Eby, et al., 2016). In summary, it can be said that older people see the driver assistance systems already available on the market as suitable for compensating for their age-related deficits and increasing driving comfort (Trübswetter, 2015). However, a large part of the existing empirical studies does not make an explicit separation between older and younger drivers, or does not examine older people as an explicit group (Eby, et al., 2016). The actual acceptance and use of the systems already available by this target group are therefore not yet sufficiently researched.

Despite the positive attitude, ADAS are not used to their full potential in reality. As mentioned at the beginning, the proportion of older drivers who actively use ADAS is currently comparably low. A study by Trübswetter (2015) showed that the use of ADAS among older people was almost exclusively limited to the cruise control and the acoustic parking aid. In addition, older test subjects stated that they could not accept technologies that increasingly take over the cognitive control of the driver. They claimed to remain the decision-making authority in the vehicle as long as possible (Ziefle, Pappachan, Jakobs, & Wallentowitz, 2008). The possible reasons for the low use of driver assistance systems by older people will be examined in more detail in the following section.

2 Problems of Older Drivers with ADAS and HMIs

The problems older people have using ADAS have various causes and are influenced by the person's experiences or self-efficacy, but also by the design of the systems. Studies show that there is a correlation between age and the fear of using technologies (Chung, Wang, Fulk, & McLaughlin, 2010). There is a negative correlation between the fear of using technologies and self-efficacy (Chung, Wang, Fulk, & McLaughlin, 2010), self-confidence (Lee & Coughlin, 2015) and a person's experience (Chung, Wang, Fulk, & McLaughlin, 2010; Niemelä-Nyrhinen, 2007). Self-efficacy refers to the confidence to successfully learn and use technical systems (Chung, Wang, Fulk, & McLaughlin, 2010). If more self-confidence, self-efficacy, or greater experience with technical systems is felt, there is less fear of using new technologies. Accordingly, positive experiences with the systems as well as perceived self-efficacy and self-confidence can reduce the fear of using corresponding technologies in older people.

Further usage barriers were identified within a qualitative study with drivers over 60 years old. In this, the low perceived usefulness of ADAS emerged as the main usage barrier for the respondents. Older people not only refer to the usefulness of the system itself, but also state that the systems do not provide appropriate support for them. The next most frequently mentioned usage barriers are the functional limits, i.e., the perceived limitation of the sensors, and the purchase price (Trübswetter, 2015).

Another usage barrier is user-friendliness, i.e., the operability and learnability of the user interfaces (Kim & Christiaans, 2012). Assisting technologies can only support as long as they are intuitive and consistent to operate and reliably guide people to correct and timely reactions with the output of information or warning. The design of the HMIs for ADAS is of particular relevance, as the HMI contributes significantly to the development of the mental model, which the user needs to correctly assess and learn the system behavior and system limits in the driving context (Strand, Stave, & Ihlström, 2018). The quality of an HMI as the interface between technology and human is thus a decisive factor for the positive experience and thus also the frequency of use of driver assistance systems (Harvey, Stanton, Pickering, McDonald, & Zheng, 2011). It must also be considered that a driver assistance system HMI usually also includes an important safety-relevant component, as the driver is an integral part of the system's safety concept. Through the HMI, he or she is involved in it to ensure safe use. The controllability of the vehicle must demonstrably remain secure even in the event of a fault, such as a false warning or a system failure (Knapp, Neumann, Brockmann, Walz, & Winkle, 2009).

Assistance systems can particularly support older drivers when their design is congruent with their needs and various abilities (Rogers & Fisk, 2010). To ensure a benefit for older people, their specific needs must therefore be taken into account (Sixsmith & Sixsmith, 2000). HMI can thus represent a usage barrier in itself, for example, if it is not intuitively designed or understandable or does not meet the needs of the target group. On the other hand, HMI also has the potential to reduce further usage barriers, such as the fear of technology, or the low perceived usefulness, through understandable and intuitive design. Thus, HMIs can positively influence the experience with a system and contribute to increased and safer use of the systems. The development of HMIs also involves, but is not limited to, the visual appearance of the HMIs. A study by Kim and Christiaans (2012) already showed that older people complained more often about sensory qualities of electronic devices such as shape, appearance, size, texture, noise or haptics than younger people. In addition to an appealing design, it was also important for older people that electronic devices are designed to be understandable and provide feedback (Kim & Christiaans, 2012). This has a particularly high value in the design of HMIs for driver assistance systems under the aspect of safe interaction with the system and its limitations (Knapp, Neumann, Brockmann, Walz, & Winkle, 2009). The interaction logic of the HMI is a decisive factor. Interaction logic means, for example, how the operating paths are structured, how certain functions are activated or whether the paths within an application are logical. The following will present the peculiarities of the target group of older drivers that are relevant to the HMI and explain their relevance to design.

3 Design-Relevant Characteristics of the Target Group of Older Drivers

In order to specifically introduce older people to supportive ADAS technologies and increase the frequency and safety of use, it is important to pay attention to the peculiarities that this age group brings with it. These include sensory deficits in the areas of vision and hearing, as well as physical and cognitive deficits. This section will further discuss the individual deficit areas and their relevance for the context of driving and HMI development.

3.1 Sensory Deficits

Visual Deficits

Visual perception is the most important sensory channel in the driving context. Up to 95% of driving-related content is received via the visual channel (Shinar & Schieber, 1991). This can include information outside the vehicle, such as other road users or road signs, as well as information inside the vehicle, such as display content and indicators. In older people, a regression of various driving-relevant visual functions is observed. These include static and dynamic visual acuity, the field of view, seeing in the dark, as well as light-dark adaptation, contrast sensitivity, motion vision, and color perception (Davidse, 2006; Shinar & Schieber, 1991). Visual abilities generally decline from the age of 40 and can have an impact on driving performance for the affected individuals (Trübswetter, 2015). The following section will take a closer look at the various visual abilities and discuss their relevance for HMI design.

Static and dynamic visual acuity

The decline in static visual acuity occurs continuously, with the initial decline in visual acuity starting around the age of 60. However, deficits in visual acuity can be well compensated for with glasses and contact lenses. If a person uses such a correction (glasses and contact lenses), then static visual acuity is the visual ability least affected by age-related decline (Shinar & Schieber, 1991). Dynamic visual acuity describes the ability to recognize details of moving objects. In the context of driving, a decline in dynamic visual acuity is most highly correlated with the risk of being involved in an accident (Shinar & Schieber, 1991). Its age-related decline begins earlier than the decline in static visual acuity. Dynamic visual acuity decreases from the age of 50 (Shinar & Schieber, 1991). The perception of moving objects also plays a crucial role in the consideration of displays. Thus, a decline in dynamic visual acuity may be associated with poorer perception of moving HMI components. On the other hand, displays also offer the opportunity to highlight moving, relevant objects from the driver's environment, thereby improving their perception.

Seeing in the dark

The ability to see in the dark also decreases with age. A study showed that the distance at which a traffic sign could still be clearly recognized at night was about 30% shorter for drivers over 60 years of age than for drivers under 25 years of age. The deterioration of vision in low light conditions not only begins earlier but is also overall greater than the decline in general visual acuity (Shinar & Schieber, 1991). In addition, the light-dark adaptation in older people is substantially slower compared to younger people. For example, it was shown that the takeover of the rod system, which is responsible for seeing in the dark, was slowed by up to two and a half minutes in subjects over 70 years of age compared to 20- to 30-year-olds (Jackson, Owsley, & McGwin Jr., 1999). Also relevant for driving in low light conditions is seeing during and after glare. Glare can occur while driving due to external static light sources, other road users, or even the displays in the vehicle. Older people show lower visual performance during glare than younger people. They also take longer to recover from glare. For the oldest subjects (over 85 years), it took on average one minute longer (1.20 min.) to recover from a one-minute glare than for participants between 65 and 74 years of age (20 s.). In addition, there was greater variability in recovery time after glare for older people, so that 25% of people in the oldest group even took longer than 3 min to recover from a one-minute glare (Haegerstrom-Portnoy, Schneck, & Brabyn, 1999). Light-dark adaptation as well as seeing during and after glare are particularly relevant in the HMI context, as drivers sometimes look from the bright HMI to the dark road during night drives and their eyes must therefore quickly adapt to the different light conditions. In addition, elements of the environment that are difficult to recognize can be highlighted by the HMI.

Field of View

The field of view, that is, the area in which one can see sharply and fix objects with the eyes, plays a crucial role for visual functionality. Studies have already shown that there is an age-related reduction in the field of view. A study by (Ramrattan et al. 2001) showed that 3% of the study participants between 55–64 years were affected by a loss of the field of view. Among the participants over 85, 17% were affected (Ramrattan, et al., 2001). In addition to the central field of view, peripheral vision also plays an important role in the driving context. In peripheral vision, objects are initially roughly recognized in order to subsequently shift the gaze to the corresponding objects. The ability for peripheral vision also decreases with age (Cohen 2008). There is particularly high variability among participants in old age. Some participants were able to perceive visual stimuli in the peripheral field of view well, while others showed no perception in the peripheral field when they focused on a stimulus in the central visual field (Haegerstrom-Portnoy, Schneck, & Brabyn, 1999). A study in which test subjects from different age groups were to recognize stimuli that migrated from the periphery to the center of the field of view also showed a reduction in the peripheral field of view. In the group of 60- to 69-year-olds, 45% of the subjects were able to capture less than 30° of the field of view, in the group of over seventies it was 65%. The resulting restriction of vision in the driving context led, among other things, to oncoming vehicles only being clearly recognized from a distance of 50 meters (Isler, Parsonson, & Hansson, 1997). The reduction of the field of view presents the HMI with the challenge of displaying relevant information in the person's central field of view. But this also offers the opportunity for the HMI to move information from the environment into the central field of view through display elements. An example of this is the blind spot assistant already available on the market. This usually displays information from the blind spot in the corresponding exterior mirror (Eby, et al., 2016) and thus moves it closer to the driver's field of view.

Contrast Sensitivity and Color Perception

Contrast sensitivity refers to the ability to see target objects with low contrasts sharply. Various studies have shown a decline in the contrast sensitivity of older test subjects (Shinar & Schieber 1991). Contrast sensitivity does not seem to decline significantly until the age of 65, but then shows a strong decline. The difference in visual acuity between high and low contrasts is also greater for older people (Haegerstrom-Portnoy, Schneck, & Brabyn, 1999). Color perception also does not initially seem to be significantly affected by an age-related decline. In one study, only test subjects over 70 years of age showed difficulties in a color perception test. The difficulty that was particularly evident in these cases was the distinction between blue and yellow (Haegerstrom-Portnoy, Schneck, & Brabyn, 1999). Designing the HMI so that contrasts and different colors are easily perceptible is already recorded in different design guidelines such as DIN EN ISO 9241 (e.g.: DIN EN ISO 9241–302, and – 303) (DIN German Institute for Standardization e. V. 2012; DIN German Institute for Standardization e. V 2009). However, this principle must be taken into account more strongly when designing for older drivers. For example, the guideline SAE J2217, which is used in the automotive industry among others, describes that display colors must be clear and simple and that lower display brightness should be used to better meet the needs of older drivers (Young, Koppel, & Charlton, 2017).

Auditory Deficits

In addition to visual stimuli, acoustic stimuli also play a central role in the driving context. A multitude of automobile manufacturers already use auditory displays in their vehicles. Auditory displays use sounds or speech in user interfaces (Frauenberger & Stockman, 2009). An example of this are acoustic parking aids, which alert the driver through sounds that they are approaching an obstacle. Various driver assistance systems use acoustic warnings to attract the driver's attention and convey information. The advantage of acoustic information is that it is independent of the line of sight (Bazilinskyy, Petermeijer, Petrovych, Dodou, & de Winter, 2018). Acoustic warnings are therefore particularly suitable when the visual information could be outside the driver's field of vision. The ability to hear, i.e., to perceive acoustic stimuli, poses a particular challenge in the driving context. The environment of a driver is very restless and loud. One must pay attention to numerous vehicle-specific auditory warning signals as well as external auditory signals, such as honking or the sirens of emergency vehicles (Amira, Irato, Zanela, Brescia, & Turki, 2018). Just like visual abilities, auditory abilities are also affected by age-related decline. For example, there are deficits in hearing higher frequencies and the spatial assignment of tones is impaired (Davidse, 2006). Despite possible impairments in hearing performance, studies have shown that older drivers find acoustic warnings in the car acceptable and desirable. A study on a junction assistant also showed preferences of older participants for an acoustic HMI over a visual HMI (Ziefle, Pappachan, Jakobs, & Wallentowitz, 2008). Auditory displays, which are adapted to the needs of older drivers, therefore have the potential to compensate for visual weaknesses of the users and support them in the use of assistance systems. Particularly multimodal displays, which use visual and acoustic output, can compensate for the respective weaknesses (visual/auditory) and thus lead to an improved perception of relevant information.

3.2 Physical Deficits

With increasing age, strength and mobility deteriorate. This includes reduced neck mobility (Isler, Parsonson, & Hansson, 1997); movement speed (Desrosiers, Hérbert, Bravo, & Rochette, 1999; Desrosiers, Hérbert, Bravo, & Dutil, 1995), as well as dexterity and coordination (Desrosiers, Hérbert, Bravo, & Rochette, 1999; Clark, Loftus, & Hammpond, Clark, 2011; Shinohara, Kang, Zatsiorsky, & Latash, 2003; Cole, 2006; Walston, 2012; Hughes, Frontera, Roubenoff, Evans, & Singh, 2002) The physical deficits are relevant in the context of driving, as they lead to vehicles being overlooked in the blind spot, or difficulties in operating the dashboard (Davidse, 2006). The following therefore describes the physical difficulties of older people in the context of driving.

Movement Speed

The speed of executing movements decreases with age. In a laboratory study, the movements of older people were slower and the accuracy was lower than that of the younger participants. When they had to perform a task with fast movements, the time they needed differed noticeably from that of the younger participants (Desrosiers, Hérbert, Bravo, & Dutil, 1995). Older people also tend to use compensation strategies to reduce their number of errors. For example, they drive slower to counteract their impaired mobility and make better decisions. However, accident statistics show that older drivers are involved in accidents above average due to their deteriorated perception and motor skills (Raw, Kountouriotis, Mon-Williams, & Wilkie, 2012), which could not be compensated by slower driving. The reduced movement speed requires an age-appropriate design of all time-limited HMI elements, such as time-outs (automatic closing) of offers and notifications. In addition, the timing of the warning (Wilschut, et al., 2014) about a dangerous situation can play a crucial role in the design of age-appropriate driver assistance systems.

Neck Mobility

As already described in the section on sensory deficits, the central and peripheral field of vision decrease with age. In addition to the decline in visual abilities, the reduced neck mobility of older people is another relevant factor (Davidse, 2006). The maximum possible rotation of the head to the left and right is between 80° and 90° for younger people. For those aged 60 to 69, it is between 60° and 70°, and for those over 70, it is under approximately 60°. This has consequences for the ability to see (Isler, Parsonson, & Hansson, 1997), as not only the field of vision itself decreases, but also the ability to shift the field of vision by turning the head.

Dexterity and Coordination

A study conducted over 3 years found a decline in various motor skills. For example, fine and gross motor skills, motor coordination, or the strength of the participants' grip were affected by age-related degradation (Desrosiers, Hérbert, Bravo, & Rochette, 1999). In addition, finger dexterity (Clark, Loftus, & Hammpond, 2011) and coordination (Shinohara, Kang, Zatsiorsky, & Latash, 2003) decrease with age. For example, a study comparing the accuracy of finger movements of older and younger subjects found that younger people can press a button more precisely in a certain direction (Cole, 2006). In addition, older people can exert less force with their fingers than younger people (Shinohara, Kang, Zatsiorsky, & Latash, 2003). This may be due, among other things, to the fact that with aging, muscle mass is increasingly less renewed and muscle strength decreases (Walston, 2012; Hughes, Frontera, Roubenoff, Evans, & Singh, 2002). Therefore, it is to be expected that older people will have increased difficulties with operating devices in the car. Coordination and dexterity play a particularly important role in the driving context, especially in the operation of the user interface. For example, many and relatively small buttons must be used without error. Through various operating concepts, such as touchscreens, buttons can be displayed larger or the required force to press them can be reduced. Thus, age-related deficits in motor skills can be compensated by the HMI.

3.3 Cognitive Deficits

In addition to sensory and physical problems, deficits in the area of cognitive abilities also occur with increasing age. Cognitive abilities such as attention, the ability to multitask, reaction speed, and memory are affected by age-related decline (Kok, 2000; Alm & Nilsson, 1995; Strayer & Drews, 2007; Muller & Weinberg, 2011) and are associated with driving performance and the risk of being involved in an accident (Anstey, Wood, Lord, & Walker, 2005). Specifically in the driving context, these deficits are reflected in the increased error rate in decision-making processes in complex driving situations (Rakontonitainy & Stainhardt 2009, September). In addition, older drivers tend to be more distracted when performing more complex tasks in the driving context Kim & Son (2011). Therefore, these deficits will be examined in more detail in the following section, particularly considering their HMI relevance.

Attention

Kok (2000) analyzed in a meta-analysis the EEG data of various studies on different types of attention such as focused, involuntary, and divided attention. Focused attention deals with the question to what extent individuals can focus their attention on a specific part of information or inputs without being distracted by irrelevant input. The meta-analysis of EEG data shows that older people are more often distracted by irrelevant peripheral stimuli that are presented simultaneously with the target stimulus (Kok, 2000). Involuntary attention is when the focus is almost automatically directed to very different, important or intense stimuli (Kok, 2000). This function of attention plays a role in the HMI context, for example, when sudden warning notices are presented that require attention. The studies included in the meta-analysis show that older people increasingly fail to recognize whether a presented stimulus differs from the background (Kok, 2000). This probably leads to a lower involuntary attention being directed to this stimulus. Divided attention involves the ability to control attention in such a way that two tasks can be performed simultaneously. To do this, attention must be directed simultaneously to different inputs or memory elements. If this is not successful, interference can occur, meaning that tasks or information overlap and therefore cannot be properly executed or perceived. During the driving task, there are more situations in which drivers have to perform several tasks at the same time (Strayer & Drews 2007) or deal with multimodal input (Muller & Weinberg 2011). This not only involves display content, but also input generated by external vehicle devices. For example, the use of mobile phones can significantly worsen the reaction time, especially in older people (Alm & Nilsson 1995). When drivers interact with devices within the car, interference can therefore occur (Salvucci, 2002). However, it has been shown that there are differences depending on the type of tasks people perform while driving. Visual secondary tasks had a greater negative effect than auditory secondary tasks. This was the case in all age groups, but was even more pronounced for older people (Wood, Chaparro, & Carberry, 2007). The studies included in Kok's (2000) meta-analysis also showed that older participants were slower and made more errors in complex visual and memory search tasks than younger ones. The optimal use of attention resources plays a crucial role in the driving context at various points. For example, drivers must pay attention to relevant information while ignoring irrelevant information in mostly complex situations (Anstey, Wood, Lord, & Walker, 2005). For example, they must focus on a difficult traffic situation and not be distracted by stimuli irrelevant to this situation. There is also a correlation between the ability to pay attention and the risk of being involved in accidents (Anstey, Wood, Lord, & Walker, 2005). To avoid interference, it may also be useful to use multimodal interfaces so that, for example, a symbol on the display and a warning tone can be processed simultaneously. Accordingly, the HMI of an assistance system should save attention resources and not take too much attention away from the driving task. At the same time, however, it also offers the opportunity to attract the attention of the driver and specifically direct it to important information.

Reaction Time

A meta-analysis by Anstey et al. (2005) revealed a correlation between reaction time and driving ability. This is particularly true for complex reaction time (Anstey, Wood, Lord, & Walker, 2005), which can also be found in the driving context. Crook, West & Larrabee (1993) examined age-related changes in reaction time in a simulated driving situation. Here, slower reaction times were shown in older test subjects, which were more influenced by cognition than by psychomotor skills. More specifically, this means that the duration of the reaction time depends more on the processing of visual input and the decision to move the hand than on the duration of the switch between the gas and brake pedal (Crook, West, & Larrabee, 1993). As already mentioned in connection with the decline in movement speed, the timing of a warning, for example, plays an important role in the age-appropriate design of driver assistance systems (Wilschut, et al., 2014). An early warning can significantly extend the available reaction time before a dangerous situation and thus lead to increased traffic safety, especially for older drivers.

Learning Ability and Memory

In addition to attention and reaction time, memory is also affected by age-related decline (Spencer & Naftali 1995). This includes, among other things, the ability to recognize already learned content. However, the free recall of learned content is even more affected by a decline (Spencer & Naftali 1995; Nagesharo, Tseng, & Filev, 2019). Age-related weaknesses in memory occur more frequently than, for example, physical problems (Nagesharo, Tseng, & Filev, 2019, October). Memory plays an important role in the context of driving. For example, a study with older test subjects showed that performance in a memory task correlated with general driving ability (Freund, Gravenstein, Ferris, Burke, & Shaheen, 2005). Memory, for example, takes on the task of remembering certain places along the route (Spencer & Naftali, 1995). To accommodate the memory and learning ability of older people, operating modes, feedback, and instructions should be adapted to their abilities (Mynatt & Rogers, 2001). It makes sense to use clear icons and earcons so that their meaning is understandable in itself and does not have to be retrieved from memory. In addition, the HMI can serve as a memory aid by making important information available again after some time.

With regard to all the age-related deficits mentioned before, it should be noted that people of the same age differ greatly in the impairment of their sensory, physical, or cognitive abilities. In some cases, such driving-relevant impairments also occur at a younger age, as a report by TÜV Kraftfahrt GmbH shows (Poschadel, et al., 2012). For this reason, HMI components should, if possible, be able to be flexibly adapted to the needs of the individual user. For example, the brightness of the display or the volume of warning tones could be individually adjustable, thus achieving the best possible fit to the various needs.

Presentation of specific HMI variants.

As already described, age-related decline in sensory, physical, and cognitive abilities presents drivers with different challenges. The deficits lead to increased demands on users in the driving task and in the operation of driver assistance systems. There are different ways to counteract these deficits. For example, technical improvements in sensors, computing power, and automation can lead to future driver assistance systems relieving and supporting drivers even more efficiently. However, improved technology only brings benefits if it is also used. At this point, the development of age-appropriate HMIs for driver assistance systems faces a special challenge. The user-friendliness of a system can influence older people to use the systems more frequently (Ziefle, Pappachan, Jakobs, & Wallentowitz, 2008). Age-appropriate HMI should not in itself present a usage barrier, i.e., it should be intuitive and easy to use. On the other hand, it has the potential to reduce further usage barriers such as the fear of using new technologies (Chung, Wang, Fulk, & McLaughlin, 2010) or the low perceived usefulness (Trübswetter, 2015). In addition, an age-appropriate HMI contributes to reducing the demands on drivers and thus enabling safer mobility for this target group. For this reason, we want to derive potential HMI solutions from the design-relevant deficits already listed and present them using specific examples. We will not only focus on technology already available in vehicles, but also include new technologies with great potential, such as an Augmented Reality (AR) windshield, in our examples.

3.4 HMI Concepts 1 and 2: Forming an Emergency Corridor and Right-of-Way Warning in the Windshield

Head-Up Displays (HUD) are already available in many vehicles. They offer potential support by allowing drivers to keep their eyes on the road while perceiving display content. HUDs will become larger and more powerful in the future, opening up many new usage possibilities (Bentacur, Villa-Espinal, Osorio-Gómez, Cuéllar, & Suárez, 2018). An ideal scenario includes an augmented display in the windshield, which will be increasingly available in vehicles in the future. It offers new options for displaying information from the vehicle's surroundings in the driver's field of vision. The concepts presented here showcase particularly relevant and promising uses of such an AR display in the windshield (Fig. 1).

Fig. 1 Note: 'Forming an emergency corridor' in the windshield

HMI Concept 1 "Note: 'Form a rescue lane' in the windshield"

In this concept, the information of an approaching emergency vehicle with the note to form an emergency corridor is displayed in the windshield. This concept has particular potential to support older drivers as it can compensate for various age-related deficits. On the one hand, all information is displayed in the central field of vision, taking into account the smaller field of vision in old age and eliminating the need for additional neck movement. On the other hand, this central and visual display compensates for possible deficits in the hearing ability of drivers. The note to form an emergency corridor can also reduce reaction time as it provides a direct and clear instruction, saving cognitive resources in case of doubt (Fig. 2).

HMI Concept 2 "Right-of-way warning in the windshield"

This concept facilitates the driver's perception of the vehicle with the right of way, as it is clearly highlighted. It also saves attention resources. Thus, the driver has to process less information from different places, but directly receives the relevant information (which right-of-way situation exists here) at the relevant place (which road user accordingly has the right of way) (Fig. 3).

Fig. 2 Right-of-way warning in the windshield

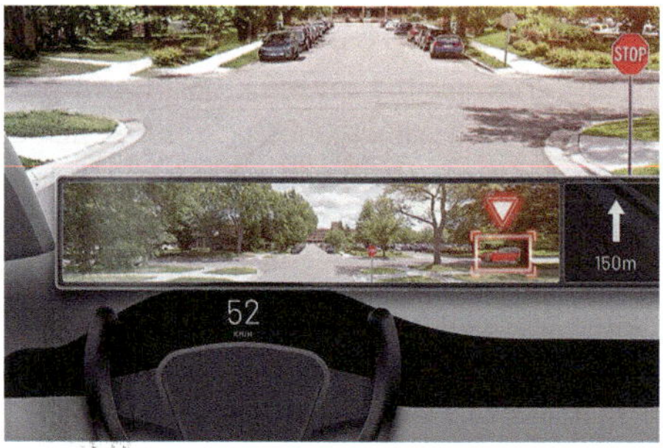

Fig. 3 Widescreen camera image in an HHD display

3.5 HMI Concept 3: Widescreen—Camera Image in an HHD Display

In addition to a HUD, a High—Head—Down—(HHD-) Display also offers a good opportunity to bring information into the driver's central field of vision. These displays are expected to be significantly larger and with increased resolution in vehicles in the future. Good examples of this display trend are the currently emerging models Mercedes-Benz EQC (twice 10.3 inches), Honda E (twice 12.3 inches), or Byton M-Byte (once 48 inches). A study by BASt, in which the "field of vision assistant" was evaluated, already provides an interesting example of the usefulness of displaying a warning in the central field of vision. A warning was displayed in the central display when a vehicle approached the intersection but was still in the peripheral field of vision (Kupschick, Bürgelen, Jürgensohn, & Protzak, 2019). The wide-angle HHD display shown here complements such a warning in the central display with a wide-angle representation of the intersection situation in the driver's central field of vision. In this example, another area of an intersection is displayed compressed in the HHD display. In addition, the approaching road user is highlighted in color and by a warning sign. This can compensate for age-related weaknesses, such as the restricted field of vision or reduced neck mobility. The early warning of an approaching vehicle can also compensate for the reduced reaction time and draw attention to the approaching vehicle. This can support drivers, for example, in particularly dangerous intersection situations for them (Fornells, Parera, Ferrer, & Fiorentino, 2017). To avoid overstimulating the attention of individuals, the age-related deficits already discussed should be taken into account when designing such a display.

3.6 HMI Concept 4: Avatar Interaction

A usage barrier in old age can be the general fear of technology (Chung, Wang, Fulk, & McLaughlin, 2010). Here, an avatar modeled after an experienced passenger can support drivers by introducing and explaining important functions and operations in calm situations. It can also overcome the barrier of low perceived usefulness (Trübswetter, 2015) by pointing out useful features or usage possibilities to the driver and explaining them in more detail. Practically, this could mean offering the driver to activate the traffic jam assistant in a traffic jam situation or describing the range of the assistance system (Fig. 4).

An avatar can also reduce the cognitive demands on drivers and relieve memory by providing important information again. Interaction with the avatar through Natural Language Understanding also allows the user to focus visual attention on the driving task while he or she is informed about an assistance system. Motor problems in operating buttons can also be compensated in this way. When designing the avatar, the described visual and acoustic deficits should be taken into account in order to design it as age-appropriate as possible.

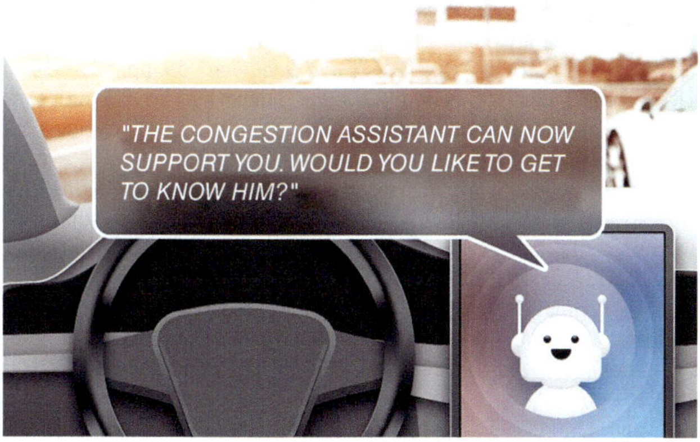

Fig. 4 Avatar Interaction

4 Summary and Outlook

Driver assistance systems already available on the market offer great support potential for older drivers and upcoming technology promises increasingly safe mobility for this group of people. However, the systems must also be usable for older people. This can be supported by an age-appropriate HMI. This HMI must take into account the peculiarities of the target group in the areas of sensory, motor and cognitive skills and compensate for the weaknesses of the respective person. In this way, it can enable a positive experience with the system and overcome usage barriers such as fear of technology or a low perceived usefulness of the systems. Therefore, in the future, the target group of older users should be more in focus and systems should also be tailored to them. Both existing and future technology such as an augmented windshield or an intelligent avatar offer promising possibilities for designing user interfaces in an age-appropriate manner. In this way, the potential of driver assistance systems for the target group of older people can be better exploited and safe mobility can be enabled.

References

Alm, H., & Nilsson, L. (October 1995). The effects of a mobile telephone task on driver behaviour in a car following situation. *Accident Analysis & Prevention, 27*(5), pp. 707–715.

Amira, B., Irato, G., Zanela, A., Brescia, A., & Turki, M. (November 2018). Congruent auditory display and confusion in sound localization: Case of elderly drivers. *Transportation Research Part F: Traffic Psychology and Behaviour, 59*, pp. 524–534.

Anstey, K. J., Wood, J., Lord, S., & Walker, J. G. (January 2005). Cognitive, sensory and physical factors enabling driving safety in older adults. (Elsevier, Eds.) *Clinical psychology review, 25*(1), pp. 45–65.

Bazilinskyy, P., Petermeijer, S. M., Petrovych, V., Dodou, D., & de Winter, J. C. (April 2018). Take-over requests in highly automated driving: A crowdsourcing survey on auditory, vibrotactile, and visual displays. *Transportation research part F: traffic psychology and behaviour, 56*, pp. 82–98.

Bentacur, J., Villa-Espinal, J., Osorio-Gómez, G., Cuéllar, S., & Suárez, D. (September 2018). Research topics and implementation trends on automotive head-up display systems. *International Journal on Interactive Design and Manufacturing (IJIDeM), 12*, pp. 199–214.

Chung, J. E., Wang, H., Fulk, J., & McLaughlin, M. (November 2010). Age differences in perceptions of online community participation among non-users: An extension of the Technology Acceptance Model. *Computers in Human Behavior, 26*(6), pp. 1674–1684.

Clark, J., Loftus, A., & Hammpond, G. (July 2011). Age-related changes in short-interval intracortical facilitation and dexterity. *Neuroreport, 22*(10), pp. 499–503.

Cohen, A. S. (2008). Wahrnehmung als Grundlage der Verkehrsorientierung bei nachlassender Sensorik während der Alterung. (T. M. GmBH, Eds.) *Leistungsfähigkeit und Mobilität im Alter*, pp. 1837.

Cole, K. J. (December 2006). Age-related directional bias of fingertip force. *Experimental Brain Research, 175*(2), pp. 285–291.

Crook, T. H., West, R. L., & Larrabee, G. J. (November 1993). The driving-reaction time test: Assessing age declines in dual-task performance. *Developmental Neuropsychology, 9*(1), pp. 31–39.

Crump, C., Cades, D., Lester, B., Reed, S., Barakat, B., Milan, L., & Young, D. (2016). Differing perceptions of advanced driver assistance systems (ADAS). *Proc. of the Human Factors and Ergonomics Society Annual Meeting, 60*, pp. 861–865. Sage CA: Los Angeles, CA: SAGE Publications.

Davidse, R. (2006). Older drivers and ADAS: which systems improve road safety? *IATSS research, 30*(1), pp. 6–20.

DESA, UN. (2019). *World Population Prospects 2019: Highlights.* Retrieved February 25, 2020, from https://population.un.org/wpp/Publications/Files/WPP2019_Highlights.pdf

Desrosiers, J., Hébert, R., Bravo, G., & Rochette, A. (1999). Age-related changes in upper extremity performance of elderly people: a longitudinal study. (Elsevier, Eds.) *Experimental gerontology, 34*, pp. 393–405.

Desrosiers, J., Hérbert, R., Bravo, G., & Dutil, É. (1995). Upper-extremity motor co-ordination of healthy elderly people. *Age and Ageing, 24*(2), pp. 108–112.

Deutscher Verkehrssicherheitsrat e.V. (2006). *Fahrerassistenzsysteme – Innovationen im Dienste der Sicherheit.* Bonn, Deutschland.

DIN Deutsches Institut für Normierung e. V. (2009). DIN EN ISO 9241-302 Ergonomie der Mensch-System-Interaktion – Teil 302: Terminologie für elektronische optische Anzeigen (ISO 9241-302:2009-06).

DIN Deutsches Institut für Normierung e. V. (2012). DIN EN ISO 9241-303 Ergonomie der Mensch-System-Interaktion – Teil 303: Anforderungen an elektronische optische Anzeigen (ISO 9241-303:2012-303).

Eby, D. W., Molnar, L. J., Zhang, L., Louis, R. M., Zanier, N., Kostyniuk, L. P., & Stanciu, S. (2016). Use, perceptions, and benefits of automotive technologies among aging drivers. *Injury epidemiology, 3*(1), p. 28.

Fornells, A., Parera, N., Ferrer, A., & Fiorentino, A. (2017). Senior Drivers, Bicyclists and Pedestrian Behavior Related with Traffic Accidents and Injuries. *SAE Technical Paper 2017-01-1397*, p. 11.

Frauenberger, C., & Stockman, T. (November 2009). Auditory display design—an investigation of a design pattern approach. (Elsevier, Eds.) *International Journal of Human-Computer Studies, 679*(11), pp. 907–922.

Freund, B., Gravenstein, S., Ferris, R., Burke, B. L., & Shaheen, E. (March 2005). Drawing clocks and driving cars. *Journal of General Internal Medicine, 20*(3), pp. 240–244.

Haegerstrom-Portnoy, G., Schneck, M. E., & Brabyn, J. A. (April 1999). Seeing into old age: vision function beyond acuity. *Optometry and Vision Science, 76*(3), pp. 141–158.

Harvey, C., Stanton, N. A., Pickering, C. A., McDonald, M., & Zheng, P. (2011). Context of use as a factor in determining the usability of in-vehicle devices. *Theoretical issues in ergonomics science, 12*(4), pp. 318–338.

Haustein, T., Mischke, J., Schönfeeld, F., & Willand, I. (Juli 2016). *Ältere Menschen in Deutschland und der EU.* Statistisches Bundesamt. Abgerufen am 20. 01 2020 von https://www.bmfsfj.de/blob/93214/95d5fc19e3791f90f8d582d61b13a95e/aeltere-menschen-deutschland-eu-data.pdf.

Hughes, V. A., Frontera, W. R., Roubenoff, R., Evans, W. J., & Singh, M. A. (August 2002). Longitudinal changes in body composition in older men and women: role of body weight change and physical activity. *The American journal of clinical nurtrition, 76*(2), pp. 472–481.

Isler, R. B., Parsonson, B. S., & Hansson, G. J. (November 1997). Age related effects of restricted head movements on the useful field of view of drivers. *Accident Analysis & Prevention, 29*(6), pp. 793–801.

Jackson, G. R., Owsley, C., & McGwin Jr., G. (November 1999). Aging and dark adaptation. *Vision research*, pp. 3975–3982.

Kim, C., & Christiaans, H. (2012). ‚Soft' usability problems with consumer electronics: the interaction between user characteristics and usability. *Journal of Design Research, 10*(3), pp. 223–238.

Kim, M. H., & Son, J. (April 2011). On-road assessment of in-vehicle driving workload for older drivers: Design guidelines for intelligent vehicles. *International Journal of Automotive Technology, 12*(2), pp. 265–272.

Knapp, A., Neumann, M., Brockmann, M., Walz, R., & Winkle, T. (August 2009). Code of Practice for the Design and Evaluation of ADAS. *Preventive and Active Safety Applications, eSafety for road and air transport, European Commission Project, Brüssel.*

Kok, A. (2000). Age-related changes in involuntary and voluntary attention as reflected in components of the event-related potential (ERP). *Biological Psychology, 54*(1–3), pp. 107–143.

Kupschick, S., Bürgelen, J., Jürgensohn, T., & Protzak, J. (2019). Erhöhung der Verkehrssicherheit älterer Kraftfahrer durch Verbesserung ihrer visuellen Aufmerksamkeit mittels "Sehfeldassistent". (B. f. (BASt), Eds.) *Berichte der Bundesanstalt für Straßenwesen. Unterreihe Fahrzeugtechnik, 127.*

Langford, J., & Koppel, S. (2006). Epidemiology of older driver crashes – identifying older driver risk factors and exposure patterns. *Transportation Research Part F: Traffic Psychology and Behaviour, 9*(5), pp. 309–321.

Lee, C., & Coughlin, J. F. (June 2015). PERSPECTIVE: Older adults' adoption of technology: an integrated approach to identifying determinants and barriers. *Journal of Product Innovation Management, 32*(5), pp. 747–759.

Muller, C., & Weinberg, G. (January 2011). Multimodal input in the car, today and tomorrow. *IEEE MultiMedia, 18*(1), pp. 98–103.

Mynatt, E. D., & Rogers, W. A. (December 2001). Developing technology to support the functional independence of older adults. *Ageing International, 27*(1), pp. 24–41.

Nagesharo, S., Tseng, E., & Filev, D. (October 6–9 2019). Autonomous Highway Driving using Deep Reinforment Learning. In IEEE (Eds.), *2019 IEEE International Conference on Systems, Man and Cybernetics (SMC)*, (pp. 2326–2331). Bari, Italy. Von arxiv.org abgerufen

Niemelä-Nyrhinen, J. (August 2007). Baby boom consumers and technology: shooting down stereotypes. *Journal of Consumer Marketing, 24*(5), pp. 305–312.

Poschadel, S., Falkenstein, M., Rinkenauer, G., Mendzheritskiy, G., Fimm, B., Worringer, B., . . . Rudinger, G. (2012). *Verkehrssicherheitsrelevante Leistungspotenziale, Defizite und Kompensationsmöglichkeiten äterer Autofahrer.* Bremerhaven: Wirtschaftsverlag NW: Berichte der Bundesanstalt für Straßenwesen.

Rakontonitainy, A., & Stainhardt, D. (September 21–22 2009b). In-vehicle technology functional requiremnets for older drivers. In A. Schmidt, A. K. Dey, T. Seder, & O. Juhlin (Eds.), *Proceedings of the 1st international conference on automotive user interfaces and interactive vehicular applications* (pp. 27–33). Essen, Germany: ACM.

Ramrattan, R. S., Wollfs, R. C., Panda-Jonas, S., Jonas, J. B., Bakker, D., Pols, H. A., . . . de Jong, P. (2001). Prevalence and causes of visual field loss in the elderly and associations with impairment in daily functioning: the Rotterdam Study. *Archives of Ophthalmology, 119*(12), pp. 1788–1794.

Raw, R. K., Kountouriotis, G. K., Mon-Williams, M., & Wilkie, R. M. (June 2012). Movement control in older adults: does old age mean middle of the road? *Journal of experimental psychology: human perception and performance, 38*(3), pp. 735.

Rogers, W. A., & Fisk, A. D. (September 2010). Toward a psychological science of advanced technology design for older adults. *Journals of Gerontology Series B: Psychological Science and Social Sciences, 65*(6), pp. 645–653.

Salvucci, D. D. (July 2002). Modeling driver distraction from cognitive tasks. Proceedings of the Annual Meeting of the Cognitive Science Society, 24.

Shinar, D., & Schieber, F. (October 1991). Visual requirements for safety and mobility of older drivers. *Human factors, 33*(5), pp. 507–519.

Shinohara, M. L., Kang, N., Zatsiorsky, V. M., & Latash, M. L. (January 2003). Effects of age and gender on finger coordination in MVC and submaximal force-matching tasks. *Journal of applied physiology, 94*(1), pp. 259–270.

Sixsmith, A., & Sixsmith, J. (February 2000). Smart care technologies: meeting whose needs? *Journal of Telemedicine and Telecare, 6*(1), pp. 190–192.

Spencer, W. D., & Naftali, R. (1995). Differential effects of aging on memory for content and context: a meta-analysis. *Psychology and aging, 10*(4), pp. 527–539.

Strand, N., Stave, C., & Ihlström, J. (2018, September). A case-study on drivers' mental model of partial driving automation. *25th ITS World Congress*, (pp. 17–21). Copenhagen, Denmark.

Strayer, D. L., & Drews, F. A. (2007). Multitasking in the Automobile. In *Attention: From Theory To Practice* (pp. 121–133).

Trübswetter, N. M. (2015). *Akzeptanzkriterien und Nutzungsbarrieren älterer Autofahrer im Umgang mit Fahrerassistenzsystemen.* Ph. D. Dissertation, TUM School of Education, TUM, München, Deutschland.

Walston, J. D. (June 2012). Sarcopenia in older adults. *Current opinion in rheumatology, 24*(6), pp. 623.

Wilschut, E. S., Kroon, M., E. C., de Goede, M., Cremers, A., & Hoedemaeker, M. (2014, January). The older adult road user: recommendations for driver assistance. In I. i. Gladbach (Eds.), *International interdisciplinary conference "Ageing and Safe Mobility" held at the Federal Highway Research Institute (BASt) in Bergisch Gladbach,* (pp. 1–10).

Wood, J. M., Chaparro, A., & Carberry, T. (2007). Investigation of the interaction between visual impairment and multi-tasking on driving performance. In I. J. Faulks, M. Stevenson, J. Brown, A. Porter, & J. D. Irwin, *Distracted Driving+* (pp. 623–640). Sydney: NSW: Australasian College of Road Safety.

Young, K. L., Koppel, S., & Charlton, J. L. (2017). Toward best practice in human machine interface design for older drivers: a review of current design guidelines. *Elsevier Ltd., 106*, pp. 460–467.

Ziefle, M., Pappachan, P., Jakobs, E.-M., & Wallentowitz, H. (2008). Visual and auditory interfaces of advanced driver assistant systems for older drivers. In M. K., K. J., Z. W., & K. A. (Eds.), *Computers Helping People with Special Needs. ICCHP 2008. Lecture Notes in Computer Science, vol 5105.* (pp. 62–69). Berlin, Heidelberg: Springer.

Evaluation of the Performance of Acoustic Assistance Systems for Older Drivers

Magnus Schäfer

Contents

Dr.-Ing. Magnus Schäfer, HEAD acoustics GmbH

M. Schäfer (✉)
HEAD acoustics GmbH, Herzogenrath, Germany
e-mail: magnus.schaefer@head-acoustics.de

H. Proff et al. (eds.), *Age-appropriate Driver Assistance Systems*,
https://doi.org/10.1007/978-3-658-44838-7_5

1 Introduction

Assistance systems are available in modern vehicles for a variety of situations to support drivers. These systems can be used with the aim of increasing comfort—often they are not noticeable in normal driving operation, but are particularly important in critical driving situations.

Acoustic systems offer the advantage of a high reaction speed on the one hand, and on the other hand, the perceived position of the sound source can be used as an additional level of information. For example, a cross-traffic assistant can signal from which direction cross-traffic was detected in order to direct attention. Accordingly, methods for evaluating the acoustic situation in the vehicle are necessary for an assessment of the performance of such assistance systems.

This chapter introduces a novel microphone array that records the sound field at the driver's position in the area that would also be used by a human listener for localization, followed by a brief overview of the localization of sound sources. A method is then described that performs localization of sound sources using machine learning methods. The necessary components are explained and an assessment of the performance of the method is given.

2 Localization of Sound Sources

The localization of sound sources can generally be viewed from two sides: Either the question is investigated where the sound source is actually located, this consideration is often encountered in technical issues. For example, with an appropriate localization method, the origin of an unwanted noise on a technical device could be located, which would then make constructive improvements possible. Another example would be mobile phones that use multiple microphones to amplify the user's voice signal against background noise.

However, the localization task can also be understood from the receiver's side. Here, it is irrelevant where the sound source is actually located, what is crucial is where it is perceived. In many cases, there are no major differences between the actual and the perceived source position, one of the most relevant exceptions to this is exactly the case that exists in modern cars: Playback systems with multiple speakers offer the possibility, through appropriate signal processing, to also generate so-called phantom sources (sources that are perceived at places where there are no speakers) (Pulkki, 1997).

To distinguish between the two cases, the term *location* refers to the determination of the actual position of the source and the term *localization* refers to the determination of the perceived source position.

For the evaluation of the performance of acoustic driver assistance systems, localization is the more relevant task, as the perception of the vehicle driver is the decisive criterion when using the system. However, replicating the complex human directional perception with a technical system is an extremely challenging task, for which there is still no functional approach that can adequately replicate all the properties and abilities of human listeners. Accordingly, a method is first presented that allows a location of the speakers to be carried out in a generic structure, but can easily be adapted for localization as soon as enough data on human perception is available.

3 Technical Procedures

In the localization of sound sources, so-called beamforming methods are used. They are typically model-based, i.e., they explicitly use knowledge about the propagation of acoustic signals to determine where sources are located. Among the propagation properties used, there is always at least the speed of sound, which results in differences in the runtime between the source and the various microphones. If the source is near the microphones, level differences can also be used ((Schäfer, Heese, Wernerus, & Vary, 2012), (Heese, Schäfer, Wernerus, & Vary, 2013)).

To illustrate, Fig. 1 shows a very simple example of an acoustic environment, in which there is only one sound source that emits the signal $x(k)$. In the environ-

Fig. 1 Simple example of an acoustic environment

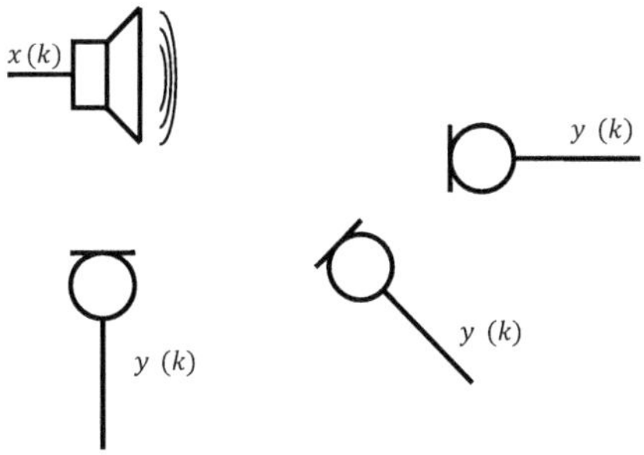

ment of the sound source, there are three microphones $M_{1...3}$, which receive the signals $y_{1...3}(k)$. The example shows an idealized free field situation, so there are no reflecting objects nearby that have an influence on the sound field.

For the situation shown in Fig. 1, the microphone signals $y_{1...3}(k)$ can be represented with distance-dependent attenuations $\alpha_{1...3}$ and delays $\tau_{1...3}$ depending on the source signal

$$y_{1...3}(k) = \alpha_{1...3} \cdot x(k - \tau_{1...3})$$

By comparing the three microphone signals with each other, runtime differences can be determined in this idealized case, from which the position of the sound source relative to the microphones can be determined. For robust localization in more complex situations with multiple sound sources or reflecting objects nearby (see the model description in (Schäfer, Multi Channel Audio Processing: Enhancement, Compression and Evaluation of Quality, 2014)), additional measures are necessary, for example, a larger number of microphones can be used, which significantly exceeds the number of possible sources, thus providing more information.

4 Human Perception

Human spatial perception has already been extensively phenomenologically investigated, an overview of this topic can be found in (Blauert, 1997). Through his two ears, man also has the ability to exploit the propagation characteristics of sound. Here, the three decisive signal characteristics can be identified, which are used for the localization of sound sources:

- Interaural level difference
  Two aspects lead to a level difference between the ears: On the one hand, one ear is closer to the sound source than the other, on the other hand, the head also has a shadowing effect. Both aspects result in the signal at the ear facing away from the source having a lower signal level.
- Interaural time difference
  The different distance to the sound source leads to a time difference in addition to the level difference: The signal reaches the ear facing away from the source later.
- Interaural cross-correlation
  As a measure of the similarity between the signals at both ears, the cross-correlation provides information about the perceived source size. A compact, point-like source is indicated by a high cross-correlation, a wide, spatially distributed source leads to a lower cross-correlation.

However, these characteristics are all not unique: A consideration of the geometric conditions shows that the characteristics are rotationally symmetric to an axis through the ears (cf. Kapralos, Jenkin, & Milios, 2008). As an illustrative example, signals from the front and from the back lead to identical level and time differences (namely no differences). To resolve this ambiguity, human listeners use small head movements, which not only analyze the sound field at two points in space, but also consider an area around the ears for localization.

5 Microphone Array Design for Perception-Correct Recording of Sound Fields

Following the approach of human listeners in the localization of sound sources, a microphone array is described below that allows the components relevant for localization to be recorded with a static measurement. A microphone array with eight microphones, specifically designed for this application, can be seen in Fig. 2, the corresponding dimensioned microphone positions are schematically represented in Fig. 3.

The microphone array can be mounted on an artificial head to also take into account the shadowing or reflecting influence of the head on the sound field

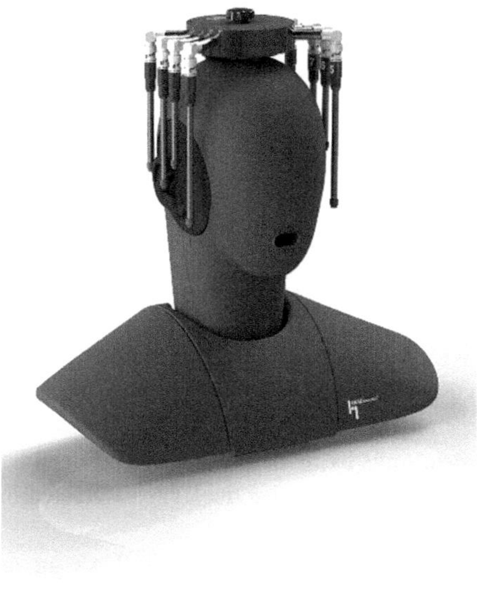

Fig. 2 Microphone array mounted on an artificial head

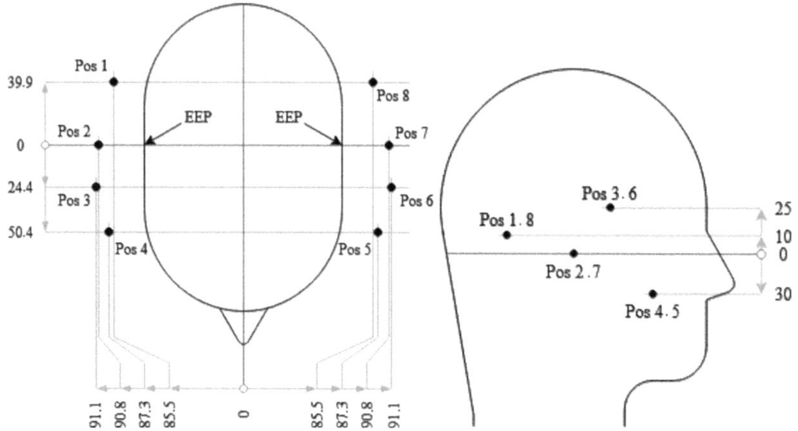

Fig. 3 Visualization of the microphone positions from above and from the side

during the recordings. The microphones are symmetrically positioned on both sides of the head and cover the typical movement space for the previously mentioned small head movements. Even though for many applications (e.g. also in the field of driver assistance systems) mainly a localization in the azimuth plane is relevant, the microphones are not positioned at the same height. The reason for this is on the one hand the possibility to also perform an estimation of the elevation of the source, but on the other hand also to cover the area that can be reached by a slight tilt of the head. It is worth mentioning the position of microphones 2 and 7, these are located directly in front of the entrance of the ear canal (ear canal entrance point—EEP) and thus provide information about the signals perceived by a human listener in a resting position.

6 Localization with Machine Learning Methods

With the microphone array, as shown, there is thus the possibility to record all components of the sound field that are also used by a human listener for localization. At the same time, the influence of the head is taken into account in the recordings by mounting it on an artificial head. The following sections describe a method that analyzes the signals obtained in this way and performs an estimation of the source position.

The localization method follows an approach based on machine learning methods, specifically neural networks. Therefore, a description of the relevant components of neural networks in general is given first, before the network structure used is introduced and explained. Subsequently, the performance of the method is evaluated using data obtained with the microphone array in a real acoustic environment. For comparison, two conventional beamforming methods are also used. In order to directly compare the results of all three methods, the data for the analysis consist of data sets in which only one source is active at a time.

7 Neural Networks

Artificial neural networks (ANN) are used in many areas of life today because of their performance and flexibility: For example, weather forecasts or automatic speech recognizers benefit from their use. The following gives a rough overview of artificial neural networks, with a particular focus on the structure of neural networks and the different types of layers.

The inspiration for artificial neural networks comes from the connections of nerve cells, for example, in the human brain. ANNs are made up of artificial neurons. Such a neuron is schematically represented in Fig. 4.

The output $\widehat{y}_k$ of the neuron can be determined depending on the input values $x_1, x_2, \ldots, x_M$ and the parameters of the neuron. The parameters of the neuron are specifically the multiplicative weighting factors $w_{k1}, w_{k2}, \ldots, w_{kM}$, the additive bias b_k and the activation function $g(\cdot)$. Overall, the following equation results for the output:

$$\widehat{y}_k = g\left(\sum_{i=1}^{M} x_i \cdot w_{ki} + b\_k\right)$$

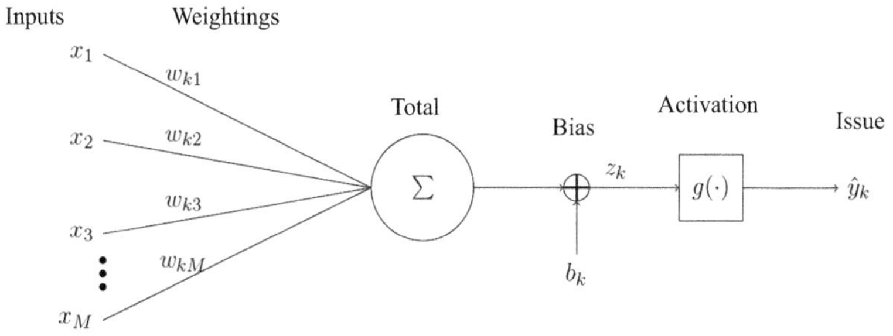

Fig. 4 An artificial neuron

The weighting factors thus represent the importance of the available inputs, and the bias represents a constant deviation that cannot be directly determined from the inputs. The activation function finally maps the sum to the value range that is necessary for the output—in the case of localization in the azimuth plane, this could be a value between 0 and 360, for example, to represent the angle in degrees.

7.1 Linking Artificial Neurons to Neural Networks

From these artificial neurons as elementary building blocks, various architectures of neural networks can be created. Here, an ANN is considered in layers—layers in which the neurons are linked in the same way. Several such layers in succession then form the neural network. The first layer is referred to as the input layer, the last layer as the output layer, and in between are the hidden layers. If there are many hidden layers, we speak of deep neural networks (LeCun, Bengio, & Hinton, Deep learning, 2015).

The following briefly introduces the different types of layers necessary for the localization system.

Fully-Connected-Layer
Fully-Connected-Layers (or Dense-Layers) are layers in which all output neurons are connected to all input nodes. In many neural networks, at least the last layer is of this type, as they have the most degrees of freedom to process the information that has been extracted throughout the network into a meaningful output.

Convolutional-Layer
Convolutional layers (LeCun, Bottou, Bengio, & Haffner, 1998) represent a method for extracting local features from the input data. They do this by applying a compact pattern of weights (filter kernel) to all areas of the input data. In a Fully-Connected Layer, each output value represents the link of an individual weight with all input data. In contrast, each output value of a Convolutional Layer is the link of a fixed weight with different parts of the input data.

By applying identical operations to small areas of the input data, a Convolutional Layer is able to identify local patterns at different locations in the input data. A chain of several Convolutional Layers is referred to as a *Convolutional Neural Network* (CNN) and can discover complex structures in the input data. Therefore, they are often used in audio and image analysis.

Pooling Layer

A Pooling Layer can be seen as a local, statistical operation on its input values. The two most common variants are Max Pooling, where the maximum value in a range of the input data is output as a result, and Average Pooling, where the result is the arithmetic mean of the input data.

A common combination is the connection of a Convolutional Layer with a subsequent Max Pooling Layer. This combination provides a data-efficient answer to the question of whether a certain feature is present at any position in the input data: The Convolutional Layer examines the complete input data for the sought-after feature. Instead of providing the analysis results for the complete data space to the following layers, the Max Pooling Layer reduces the result to one, namely the maximum value.

7.2 Structure of Neural Networks for Different Tasks

Depending on the task at hand, neural networks can be structured differently to meet the requirements of the respective task. The task considered here, the localization of a sound source, can be well represented by a chain of convolutional layers for feature extraction followed by fully connected layers for linking the information obtained (cf. (Chakrabarty & Habets, 2017)). In terms of the dimension of the output layer, two variants are basically possible: While a network for regression of a (angle) value has only a single output neuron that provides an estimate, a network used for classification into different, predefined direction classes has one output neuron per class.

8 Training of a Neural Network

Regardless of the exact structure of the network considered, the basic idea of machine learning is to determine the parameters of all artificial neurons from a set of representative training data so that the ANN also provides meaningful results for unknown data.

In training, both the input data $x_1, x_2, \ldots, x_M$ and the output data y_k are known. This allows an error e_k to be determined, which quantifies the difference between the current and the perfect output of the neuron:

$$e_k = y_k - \widehat{y}_k = y_k - g\left(\sum_{1=1}^{M} x_i \cdot w_{ki} + b_k\right)$$

This error, which is defined here for a neuron, can also be determined in a similar way for the complete ANN—the equation just becomes significantly more complex. But even then, there is a representation for the network's prediction through the combination of the input data with the network's parameters.

Typically, the available amount of data is divided: The larger part is used for training the ANN, a smaller part is used for validating the ANN, in order to be able to examine the behavior with unknown data already during the training process. In Fig. 5, the training process is shown. Training is an iterative process in which the training data is used multiple times to optimize the network parameters.

The process begins with the calculation of the output values of the current state of the neural network for the input values $\boldsymbol{x}$ from the training data. The resulting prediction $\boldsymbol{y}_k$ is then compared with the actual target values $\boldsymbol{y}_{ref}$ for these training data. This comparison is carried out with a (network structure and possibly application-specific) loss function C. The backpropagation step determines the gradient of the loss function depending on the network parameters (weights and bias). Based on this gradient, the optimizer used then modifies the parameters and another training run begins.

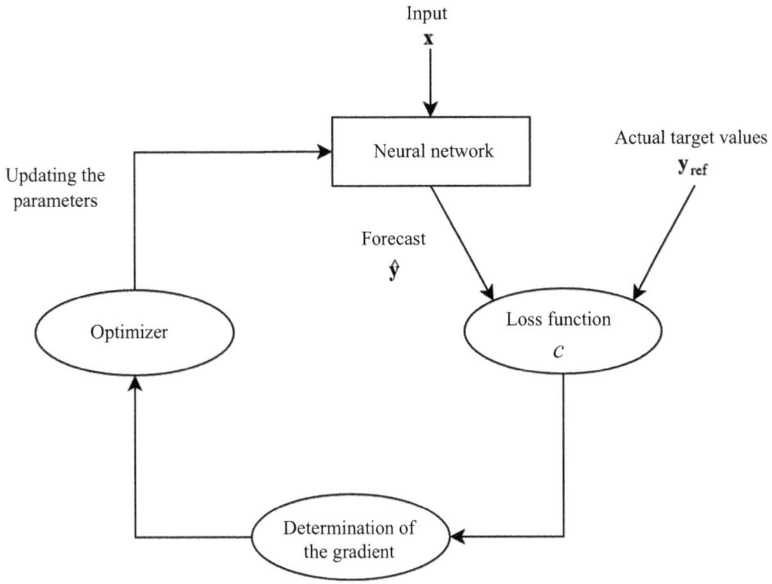

Fig. 5 Process of training an artificial neural network

The iteration loop in Fig. 5 is typically not run with the complete training data. Instead, the data is divided into several groups and an optimization step is carried out after each group. Once all training data has been processed in this way, a validation step takes place. During this, the network parameters are not updated, but only the loss function (and possibly other metrics) is calculated. A run of training and validation is referred to as an epoch.

The division into training and validation data now allows a sensible point in time to end the training to be derived from the course of the loss function (or the additional evaluation metrics) for the validation data.

8.1 Loss Functions

The choice of the loss function is a very critical element of the iteration loop for optimizing the parameters of the network. It quantifies the prediction error and ideally represents the quality of the current model. Depending on the specific task and the modeling structure used, different loss functions make sense. The following introduces two variants of loss functions (one for regression tasks, one for classification tasks).

Regression: Mean Squared Error

The mean squared error (MSE) is a classic choice for regression tasks, it is defined as (Nielsen, 2015):

$$C_{MSE} = \frac{1}{2N} \sum_{X} y_{ref}(\boldsymbol{x}) - \widehat{y}(\boldsymbol{x})^2$$

Here, N denotes the number of data sets used, $y_{ref}(\boldsymbol{x})$ are the known reference data and $\widehat{y}(\boldsymbol{x})$ is the network's prediction for the input data $\boldsymbol{x}$ from the total set $\boldsymbol{X}$. The MSE is well suited for regression tasks as it scales with the prediction error: Large deviations between prediction and actual reference value lead to large values of the loss function.

Classification: Cross-Entropy

For classification tasks, the MSE is not suitable as a loss function. Here, the cross-entropy provides a meaningful measure for evaluating the current prediction. It is defined by (Nielsen, 2015):

$$C_{CE} = -\frac{1}{N} \sum_{X} \sum_{J} \left(y_{ref,j}(x) \cdot \ln \widehat{y}_j(x) + \left(1 - y_{ref,j}(x)\right) \cdot \ln\left(1 - \widehat{y}_j(x)\right)\right)$$

Here, j is the index for the J output neurons of the KNN. As shown in (Nielsen, 2015), this loss function has the advantage that the resulting learning rate does not depend on the course of the activation function used, but only on the error in the prediction.

8.2 Optimizers

After calculating the loss function, as shown above, the gradient of the loss function is determined depending on the parameters of the network (weighting factors w_{ki} and bias values b_k). This gradient is then the input value for the parameter optimization algorithm.

The goal of this optimization is to reduce the value of the loss function. A very powerful approach for this is the *Adaptive Moment Estimation* (AdaM) (Kingma & Ba, 2014), which combines the strengths of two other methods: As with momentum optimization (Polyak, 1964), past gradients are also taken into account in the optimization and following RMSprop (Tieleman & Hinton, 2012), the adaptation speed of the network parameters is adaptively derived from the current data.

8.3 Metrics for Monitoring Prediction Accuracy

In the context of iterative optimization of network parameters, only the loss function is used. However, since this does not necessarily provide a clear representation of the current prediction accuracy, two additional metrics are also calculated from the predictions in parallel.

Mean Absolute Error

The *Mean Absolute Error* (MAE) is typically used in regression tasks, as it directly indicates the average deviation between the predicted value and the reference value, and thus directly fits the corresponding loss function:

$$MAE = \frac{1}{N} \sum_{X} \left| y_{ref}(\boldsymbol{x}) - \widehat{y}(\boldsymbol{x}) \right|$$

In the case of localization in the azimuth plane, the minimum MAE is 0° if the correct position is always determined exactly. The maximum value is 180° if the source is always determined exactly opposite the correct position. It is important to note the periodicity of the value range so that, for example, a prediction of 359° with a correct position of 0° leads to a deviation of 1° (and not 359°).

Accuracy
In the present case of localizing sound sources in the azimuth plane, an MAE can also be determined for a modeling as a classification if the predicted classes and the reference classes are converted into the corresponding angles. In the general case, this is not possible with classifications, for this the *Accuracy* (Acc) is more suitable. It is defined as the quotient of correct predictions and the total number of predictions:

$$Acc = \frac{\#\text{correct predictions}}{\#\text{predictions}}$$

8.4 Preventing Overfitting to the Training Data

A deep neural network has a very high number of degrees of freedom. Accordingly, the case can quickly arise that all peculiarities of the training data are completely captured by the DNN during the training process. This would be the ideal situation for machine learning if the training data were perfectly representative of all possible data that the DNN should analyze.

Unfortunately, it is rarely possible to generate such an extensive training dataset. Therefore, too strong an adaptation to the training data usually leads to lower performance with unknown validation or test data. To avoid this situation, it is necessary to stop the training at an appropriate time or to prevent overfitting already through components in the training.

Early Stopping
The simplest approach to avoiding overfitting is to stop training depending on the current training state (Prechelt, 1998). For this purpose, metrics are calculated for the training and validation data during the training process.

Such a metric could, for example, be the prediction error of the network. This error will continuously decrease for the training data, and a decrease in the error is usually also observed for the validation data. If the error for the validation data then rises at any point, the training is stopped.

Dropout
In addition to early termination of the learning process, there is also the possibility of intervening in the learning algorithm to prevent overfitting. One variant of

this is the integration of dropout layers (Srivastava, Hinton, Krizhevsky, Sutskever, & Salakhutdinov, 2014). A dropout layer ensures that in training, with a chosen probability, the dropout rate, different neurons are randomly not connected. However, all neurons are connected for validation. With each group of training data, a different subnet of the complete DNN is thus trained.

9 Structure of the Localization Method

The goal of the method is the localization of a sound source in terms of its azimuth angle. As already mentioned, this task can in principle be represented both in the form of a regression and as a classification. For application in the evaluation of driver assistance systems, where the acoustic signal can also be used as a control variable for the visual attention of the driver, the classification fits very well, as it is almost irrelevant whether there is a deviation of 30° or a deviation of 180°. In both cases, the target direction or the target object is not centrally in the field of view.

Accordingly, a localization procedure is described and analyzed below that combines the previously described components of neural networks into a classifier. The classifier uses 72 classes, which are evenly distributed and each represent a window of 5°.

9.1 Preprocessing of Training Data

The time signals recorded with the microphone array are first transformed into the time-frequency domain using a short-time Fourier transformation. A Hann window with an overlap of 50% and a length of 2048 samples at a sampling rate of 48 kHz is used as the window function.

The result of this transformation is three-dimensional: time, frequency, and microphone channel number. The analysis is carried out individually for each time step, leaving the two dimensions of frequency and channel numbers. To provide the convolutional layers at the beginning with well-structured data for their feature extraction, the coefficients are rearranged in each time step as shown in Fig. 6.

This arrangement approximates the geometric structure of the microphone array and thus allows the convolutional layers to relate features from channels that are in a meaningful spatial relation to each other. The feature used is the phase value of the Fourier transform. This is normalized to the value range

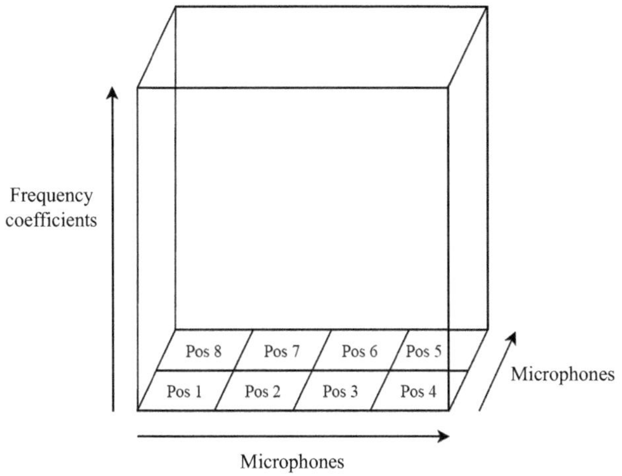

Fig. 6 Structuring of input data

between 0 and 1 before the input layer to provide a numerically favorable representation into the network.

Further preprocessing is necessary for the target values: Since the neural network is used for classification, the target values must be converted from an angle representation to a class representation. Based on the application for the analysis of driver assistance systems, the azimuth plane was divided into 72 equiangular classes. In the training data, the class in which the source is located was assigned a value of 1, all other classes a value of 0. In the application, the KNN provides with its 72 outputs the probability for each class that a source is located in the class.

9.2 Structure of the Neural Network

An overview of the used neural network is shown in Fig. 7. The analysis begins with a series of convolutional layers, which have three-dimensional or two-dimensional kernels depending on the dimensionality of the data. After these layers, the information of all microphone channels is linked. Extensive test series showed that further convolutional layers over the frequency dimensions lead to worse results in this application. To process the resulting data with the three subsequent fully connected layers, they must be reshaped into a one-dimensional tensor, which happens in the flatten layer.

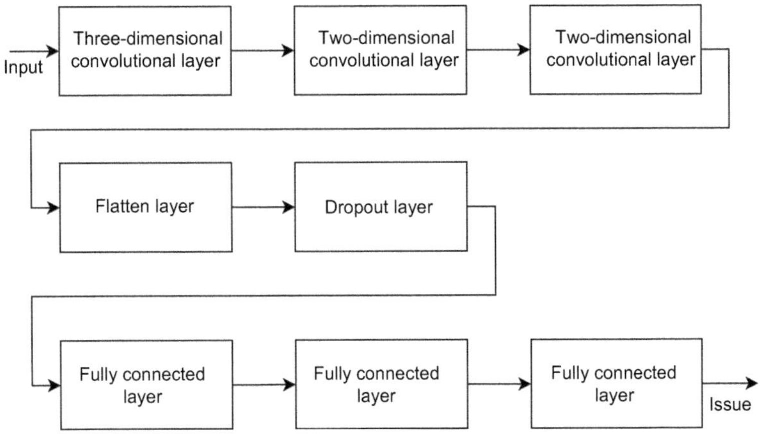

Fig. 7 Structure of the neural network

The layers used are dimensioned as follows after a hyperparameter search:

- Filter kernel of the three-dimensional convolutional layer: 256
- Filter kernel of the first two-dimensional convolutional layer: 256
- Filter kernel of the second two-dimensional convolutional layer: 64
- Dropout rate: 75%
- Neurons in the first two fully connected layers: 128
- Neurons in the last fully connected layer: 72 (corresponding to the number of classes)

For training, cross-entropy is used as the loss function, and the optimizer used is AdaM.

9.3 Training and Validation Data

For training and validation, data is used that was obtained in a real acoustic environment and not in a simulated manner. The environment consists of an acoustically treated measurement chamber (length: 3.4 m; width: 2.4 m; height: 2.02 m) with a reverberation time RT_{60} of 65.9 ms. In this measurement chamber, a total of eight speakers were mounted in the corners of the room and in the centers of the walls.

Recording of Signals

To record the signals, the previously described microphone array was mounted on a dummy head as shown in Fig. 2. This dummy head was in turn placed on a turntable to generate a dataset with as diverse source directions as possible. The speakers were at slightly different heights to also have a slight variability in the dataset. Even if it is only about estimating the azimuth angle, it cannot be assumed in the car arrangement that all speakers are exactly at the driver's ear level. The dummy head was aligned on the turntable so that its ears were approximately at the middle height of the speakers (the elevation angles range from 83.4° to 94.8°).

Two different types of signals were used: white noise, which is used for training the neural network, and speech signals for validation and for comparison with the methods from (DiBiase, 2000) and (Schmidt, 1986). The noise signals each have a length of 8 s, the speech signals (several short sentences with a length of 4 s) come from (ETSI TS 103 281 V1.3.1, 2019). The German speech signals from (ETSI TS 103 281 V1.3.1, 2019) are used for validation, the English ones as test material for the comparison of the different methods.

In the measurement process, the dummy head was rotated in 5° steps using the turntable and at each position all signals were sequentially played back from all speakers. Thus, a total of 576 spatial arrangements can be recorded in this measurement process.

Preprocessing of Data

To ensure a smooth training and validation process, the data was preprocessed to a limited extent. The noise signals were divided into segments of 1 s each and the speech signals, which were combined in a long sequence during the recordings, were divided into individual sentences. With a speech activity detection, the periods were also removed from the speech signals in which no activity is present. For the validation data, an automatic activity detection with (Wiseman, n.d.) was used for this, for the test and comparison data, passages with active speech were manually selected to exclude possible misidentifications.

Generation of Additional Data

The larger the quantity and variance of the available training data, the better the performance of the neural network is usually. One way to quickly increase the amount of data is the application of *Data Augmentation.* In this process, new data is generated from the existing data through modifications.

Adding uncorrelated noise to the training datasets is such a modification, with signal-to-noise ratios of 5 dB, 10 dB, 15 dB, 20 dB, 25 dB, 30 dB, 35 dB, and

50 dB being generated. Whether this expansion of the training data volume leads to improvements is tested by a performance comparison of two neural networks.

10 Experimental Results

As part of the experimental investigation, four different localization methods were tested:

- SRP-PHAT, the localization method from (DiBiase, 2000)
- MUSIC, the localization method from (Schmidt, 1986)
- The neural network without *Data Augmentation*
- The neural network with *Data Augmentation*

The neural networks achieved their best localization accuracies after 25 epochs (without *Data Augmentation*) and 21 epochs (with *Data Augmentation*). It should be noted that an epoch in the second case is significantly more extensive due to the much larger amount of data than an epoch in the first case.

The previously introduced metrics MAE (mean absolute error) and *Accuracy* are used as evaluation metrics. For SRP-PHAT and MUSIC, the *Accuracy* is determined by quantizing the prediction values of the two methods to the classes used for the neural networks. The mean absolute error is calculated for the neural networks by determining the deviation between the center of the chosen class and the actual source position. The values averaged over all test data are compiled in Table 1.

In the two methods from the literature, it is clear that SRP-PHAT is clearly superior to MUSIC: the mean absolute error is about 7.5° lower, the *Accuracy* about 11% higher. Comparing the two neural networks with each other, a positive influence of the *Data Augmentation* is clearly visible: The mean absolute error decreases by over 6°, the *Accuracy* increases by over 10%. If we now compare

Table 1 Evaluation of the performance of various localization algorithms

Method	Mean Absolute Error (°)	Accuracy (%)
SRP-PHAT	9.30	54.30
MUSIC	16.88	43.39
Neural Network without Data Augmentation	24.96	56.18
Neural Network with Data Augmentation	18.60	66.50

SRP-PHAT and the neural network with Data Augmentation, a mixed picture emerges: the mean absolute error of SRP-PHAT is only half as large, but the neural network has an over 12% better *Accuracy*. The neural network thus hits the exact right direction much more often, but when a misclassification occurs, the deviations from the actual source position are on average larger than with SRP-PHAT.

11 Adaptation to Perceptual Localization and Adaptability to Different Groups of People

If the method presented here is to be used for the localization of perceived source positions in more complex acoustic environments, training and validation data are necessary in which the position information is not determined from the geometric relation between microphone array and speaker position. Instead, the judgments of human listeners must be used. The effort to create a representative dataset with listening tests is high, but not utopian—not least because the *Data Augmentation* provides a tool that allows to increase the amount of data on a simulation level.

An adaptation of the properties of the method to special groups of people is possible by selecting the participants for the listening test. This can create specific analysis methods for different user groups, which can then be used, for example, for the evaluation of specific assistance systems for this user group.

12 Summary

A localization system for the evaluation of acoustic driver assistance systems based on a specially developed microphone array and neural networks was presented. The microphone array records the components of the sound field in the vicinity of the ears, which are also used by human listeners for localization. The signals obtained in this way are analyzed by a neural network, which determines the source position with a resolution of 5° through classification.

It was shown that the neural network achieves significantly higher *accuracy* compared to conventional beamforming algorithms. In addition, the method can be adapted directly to other user groups or use cases by exchanging the training and validation data.

References

Aditya, J., Bansal, R., Kumar, A., & Singh, K. (2015). A comparative study of visual and auditory reaction times on the basis of gender and physical activity levels of medical first year students. *International Journal of Applied and Basic Medical Research, 5*(2), 124–127.

Blauert, J. (1997). *Spatial Hearing: The Psychophysics of Human Sound Localization.* MIT press.

Chakrabarty, S., & Habets, E. A. (2017). Broadband DOA estimation using convolutional neural networks trained with noise signals. *IEEE Workshop on Applications of Signal Processing to Audio and Acoustics (WASPAA).* New Paltz.

DiBiase, J. H. (2000). *A High-Accuracy, Low-Latency Technique for Talker Localization in Reverberant Environments Using Microphone Arrays.* Providence: Brown University.

ETSI TS 103 281 V1.3.1. (2019). Speech and multimedia Transmission Quality (STQ); Speech quality in the presence of background noise: Objective test methods for super-wideband and fullband terminals. Sophia Antipolis: ETSI.

Fozard, J. L., Vercruyssen, M., Reynolds, S. L., Hancock, P. A., & Quilter, R. E. (1994). Age Differences and Changes in Reaction Time: The Baltimore Longitudinal Study of Aging. *Journal of Gerontology, 49*(4), 179–189.

Heese, F., Schäfer, M., Wernerus, J., & Vary, P. (2013). Numerical Near Field Optimization of a Non-Uniform Sub-band Filter-and-Sum Beamformer. *Proceedings of IEEE International Conference on Acoustics, Speech, and Signal Processing (ICASSP).* Vancouver.

Kapralos, B., Jenkin, M., & Milios, E. (2008). Virtual Audio Systems. *Presence Teleoperators & Virtual Environments, 17*, 527–549.

Kingma, D. P., & Ba, J. (2014). Adam: A method for stochastic optimization. *arXiv preprint arXiv:1412.6980.*

LeCun, Y., Bengio, Y., & Hinton, G. (2015). Deep learning. *Nature, 521*, 436–444.

LeCun, Y., Bottou, L., Bengio, Y., & Haffner, P. (1998). Gradient-based learning applied to document recognition. *Proceedings of the IEEE, 86*(11), 2278–2324.

Nielsen, M. A. (2015). *Neural Networks and Deep Learning.* San Francisco: Determination Press.

Pain, M., & Hibbs, A. (2015). Sprint starts and the minimum auditory reaction time. *Journal of Sports Science, 25*(1), 79–86.

Polyak, B. T. (1964). Some methods of speeding up the convergence of iteration methods. *USSR Computational Mathematics and Mathematical Physics, 4*(5), 791–803.

Prechelt, L. (1998). Early stopping-but when? In *Neural Networks: Tricks of the trade* (pp. 55–69). Berlin, Heidelberg: Springer-Verlag.

Pulkki, V. (1997). Virtual Sound Source Positioning Using Vector Base Amplitude Panning. *Journal of the Audio Engineering Society, 45*(6), 456–466.

Schäfer, M. (2014). *Multi Channel Audio Processing: Enhancement, Compression and Evaluation of Quality.* Aachen: Wissenschaftsverlag Mainz.

Schäfer, M., Heese, F., Wernerus, J., & Vary, P. (2012). Numerical Near Field Optimization of Weighted Delay-and-Sum Microphone Arrays. *Proceedings of International Workshop on Acoustic Signal Enhancement (IWAENC).* Aachen.

Schmidt, R. (1986). Multiple emitter location and signal parameter estimation. *IEEE transactions on antennas and propagation, 34*(3), 276–280.

Srivastava, N., Hinton, G., Krizhevsky, A., Sutskever, I., & Salakhutdinov, R. (2014). Dropout: a simple way to prevent neural networks from overfitting. *The journal of machine learning research, 15*(1), 1929–1958.

Tieleman, T., & Hinton, G. (2012). Lecture 6.5-rmsprop: Divide the gradient by a running average of its recent magnitude. *COURSERA: Neural networks for machine learning, 4*(2), 26–31.

Wiseman, J. (n. d.). *py-webrtcvad – Python interface to the WebRTC Voice Activity Detector*. Retrieved September 19, 2019, from https://github.com/wiseman/py-webrtcvad.

Driving Behavior of Older People in the Driving Simulator

Stephan Schweig and Dieter Schramm

Contents

This chapter describes the structure of the simulator used, as well as the experiments conducted. The results of the experiments are briefly summarized.

1 Simulator Setup

Driving simulators are primarily used for researching the human-vehicle interface and for investigating systems that must necessarily involve the driver in the investigation due to their functionality. Modern driving simulators enable this by creating

Dipl.-Ing. Stephan Schweig, Prof. Dr.-Ing. Dieter Schramm, all University of Duisburg-Essen

S. Schweig (✉) · D. Schramm
Lehrstuhl für Mechatronik, Universität Duisburg-Essen, Duisburg, Germany
e-mail: stephan.schweig@uni-due.de

D. Schramm
e-mail: dieter.schramm@uni-due.de

H. Proff et al. (eds.), *Age-appropriate Driver Assistance Systems*,
https://doi.org/10.1007/978-3-658-44838-7_6

the opportunity to immerse a real driver (or other people) in a realistic environment. Simulators are an versatile and popular tool, especially for the development of systems that specifically involve the driver (e.g., driver assistance). Increasingly, other issues from the field of driving dynamics or interior design are also being investigated with appropriate simulators. Driving simulators can be used wherever purely simulation technical solutions are not sufficient and driving tests in real vehicles are too expensive, too dangerous, not permitted, or if no test vehicles are available at all. In the case of the task at hand in the ALFASY project, the simulator is the only way to investigate the project-relevant issues.

When a driving simulator is used to induce realistic behavior of a driver in various scenarios, and thus to evaluate either the driver, the system, or a combination of both, a precise model for describing the physics (the correlations and behavior of various systems) is required that can be simulated in real time. In addition, the driver must be given the opportunity to immerse himself in the situation. The driver must be able to solve the tasks given (starting with the pure driving task) as he would in a real environment (Maas, Hesse, Koppers, & Schramm, 2014). These prerequisites were taken into account and met with the driving simulator used and the experimental procedures practiced.

A static driving simulator was selected for the experiment at the Chair of Mechatronics (Fig. 1). A dynamic simulator that was also available was excluded for safety reasons. The reason for this was the special drive form of the dynamic simulator. The dynamic simulator is moved by several electric linear motors. The

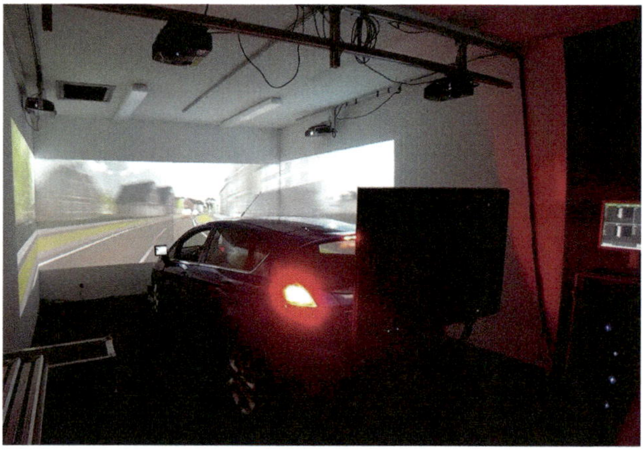

Fig. 1 Static driving simulator

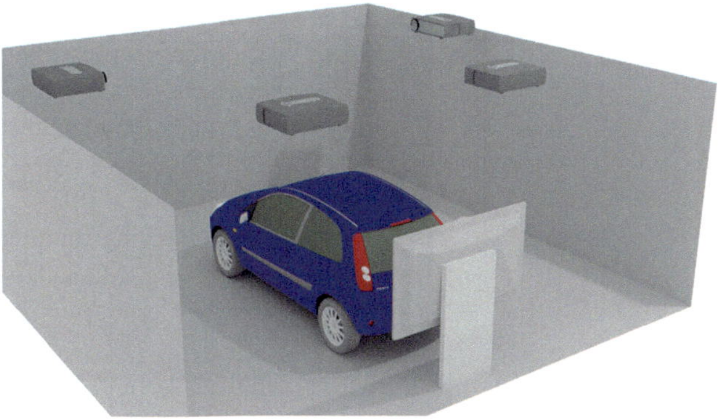

Fig. 2 Schematic representation of the simulator lab

magnetic field strength of the machines can affect smaller electronic devices in the vicinity. Therefore, entering the simulator with a pacemaker is prohibited.

All simulators and models of the chair are self-constructed, so that the complete know-how is available at the chair. Hence, the simulators can be rebuilt and reprogrammed for new experiments at any time without external help.

The basic structure of the static simulator shows Fig. 2. The vehicle used (Ford Fiesta) is located in a room whose side walls are played with projectors. These allow a field of view of around 180°. The field of view is extended by additional displays. The mirrors in the vehicle's side mirrors have been replaced by screens that also serve to visually represent the vehicle environment. In addition, a monitor of suitable width is placed behind the vehicle's rear window, which is used both for the rear view and for the reflection of the image in the interior mirror.

For the test execution, a static driving simulator was selected at the Chair of Mechatronics and supplemented with an additional touch screen and additional speakers (Fig. 3), distributed in the vehicle. In addition to the six standard speakers

Fig. 3 Speaker distribution in the simulator

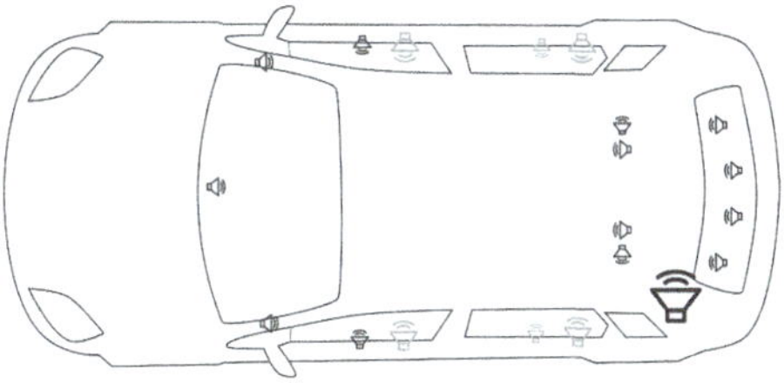

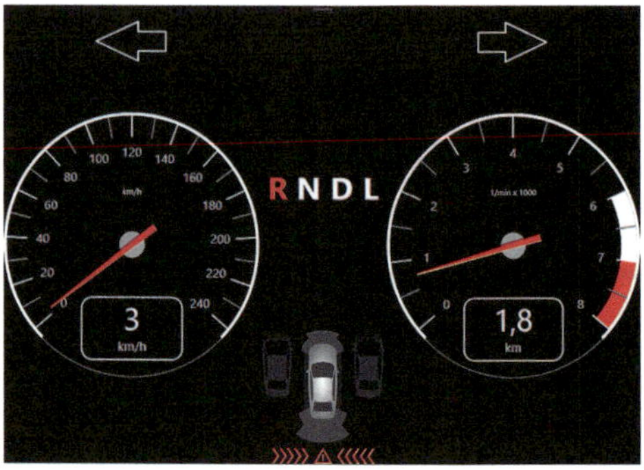

Fig. 4 Instrument cluster in the vehicle simulator

in the doors (2 front, 4 rear), additional speakers were mounted in the A-pillars, the center console, the headrests of the rear bench, and on the trunk cover.

With the exception of the steering column and analog instruments, the vehicle used is an unchanged production vehicle (Fig. 1). Instead of the instruments, a display shows speed, rpm, etc. (Fig. 4). Furthermore, depending on the driver assistance system used, a corresponding display is shown.

This allows for the adjustment of the appearance of the combination instrument to match the study, as well as to display assistance systems that were not installed in the vehicle by the manufacturer. The steering column is not connected to the front wheels and has an actuator that can imprint a travel-induced countertorque on the steering to convey a realistic steering feel. All inputs to the vehicle, such as the accelerator pedal, brake pedal, indicators, lights, gear, etc., are read directly from the vehicle network via the CAN bus and passed on to a real-time capable vehicle model. The vehicle model used represents the physical properties of the vehicle and the influence of the environment, and calculates all available vehicle parameters.

Some of the calculated values are sent to the instrument display and serve to display speed, rpm, etc. Also, some of the assistance systems send data for display to the display Fig. 5. Position and orientation in space are transmitted from the vehicle model to the graphics control for supplying the visualization systems, which assemble the image for projection.

To enable interactions in road traffic, an environmental model with additional road users is calculated and displayed in parallel on a second real-time system, Fig. 6. These data is needed by both the visualizations and the assistance systems in the vehicle model.

Fig. 5 Display of active driver assistance systems

Fig. 6 Representation of a city scene in the driving simulator (example)

The communication between the vehicle model, traffic model, and all representing systems is done via a UDP/IP network connection. The UDP protocol was chosen because it is very suitable for real-time transmissions due to its very low latency.

2 Experiments

The first trial run within the framework of ALFASY took place in a fictitious city on an area of 3×3 km. The road network of the scenario (Fig. 7) is over 70 km long and consists of various types of roads. The motorway ring around the city is connected to the city streets by country roads. The closed form of the roads in loops allows test subjects to drive indefinitely without ending up in dead ends.

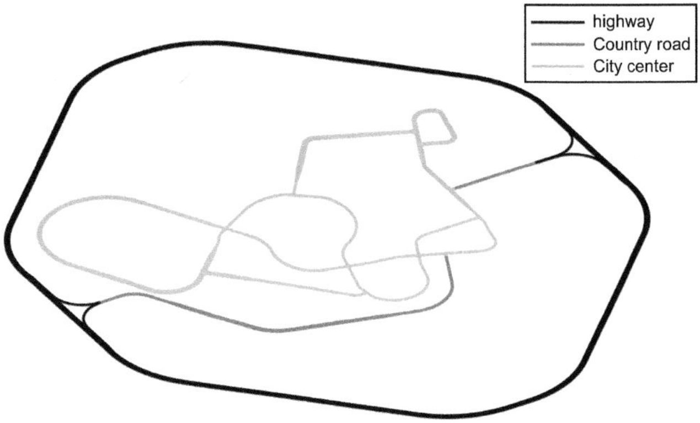

Fig. 7 City map first trial run

Before the start of the test, the participants were informed that they could terminate their participation at any time without consequences. They were then introduced to the driving simulator. In addition to technical instructions (e.g., that it is an electric vehicle), the participants were asked to choose independently within the virtual scenario how and where they drive, without following a leading car or a predetermined route. The maximum driving time in the simulator was 25 min.

For the investigation of the driver assistance systems, subjects were preselected through the first trial run. Subjects who could drive 20 min or more in the first trial—still without assistance systems—without simulator sickness occurring (see Sect. 3), were admitted to the main study with two further drives.

For the main study, which was conducted in two steps, the scenario was slightly changed. A vehicle with a combustion engine was now used as the model, which also produced the typical combustion noises. Instead of the complete city, the drive was limited to the motorway. However, two rest areas with opportunities for parallel and perpendicular parking were set up on the motorway (Fig. 8). In addition, the systems listed in Sect. 2 were activated. The systems provide visual and acoustic information.

Each trial is divided into two blocks (Fig. 9). Each block begins with a five-minute introductory drive to give the subjects the opportunity to get acquainted to the vehicle control. Afterwards, one or the other rest area is approached to park perpendicularly in block A and parallel in block B. After leaving the parking space, there are another five minutes of free driving. Both during the free driving

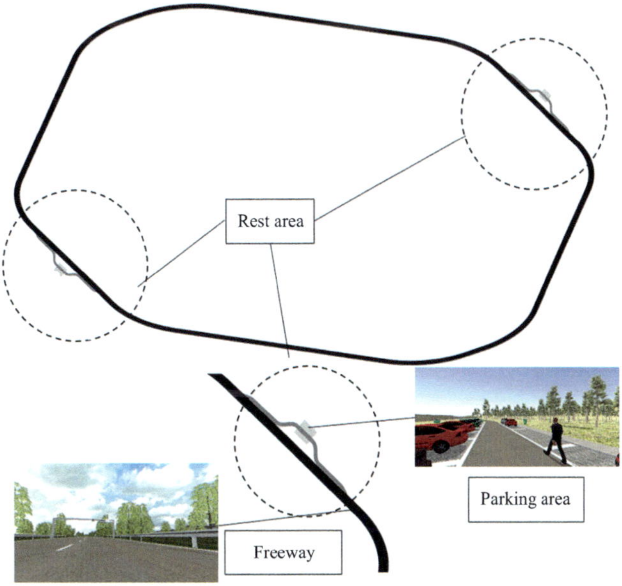

Fig. 8 Road map trial runs two and three

and during parking and unparking, situations are created that activate the used assistance systems.

The first run of the main study is identical for all subjects. The acoustic signals of the assistance systems come from the central speaker in the center console. In the second—comparative—run, the signals vary both in terms of noise and the local source, i.e., modified tones from the direction in which the event occurred. In another variant, the signals move across several sources in the room.

3 Simulator Sickness

A major challenge of all studies in simulators is the frequently occurring simulator sickness (kinetosis), which is discussed as being due to the human body's lack of adaptation to unusual combinations of sensory stimuli. This description also includes the difference, for example, between optical stimuli and detected accelerations. It is described as a phenomenon that includes symptoms such as headaches, sweating, dry mouth, drowsiness, disorientation, dizziness, nausea, and

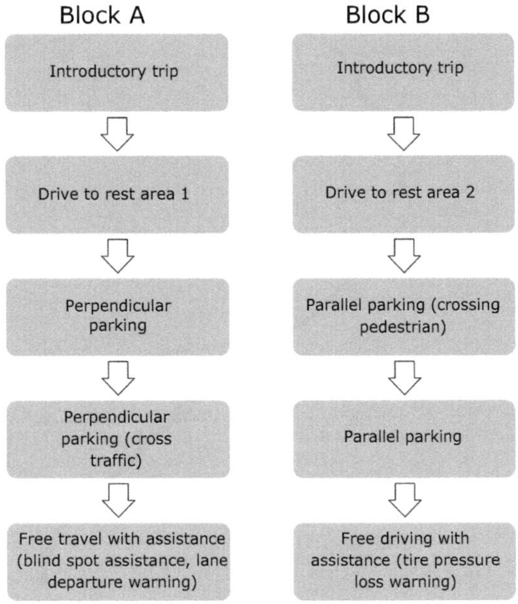

Fig. 9 Experimental procedure

vomiting (Brooks et al. 2010), which are similar to those of travel or seasickness, but typically less severe (Kennedy, Lane, Berbaum, & Lilienthal, 1993). Previous studies reported dropout rates in simulation attempts from 5% to 30% (Stanney, Kingdon, & Kennedy, 2002). Regarding gender-specific differences in the occurrence of simulator sickness, several studies (Freund & Green 2006), (Garcia, Baldwin, & Dworsky, 2010) consistently show a reduced occurrence of simulator sickness in men compared to women. In the past, numerous theories have tried to provide a better understanding of the cause of simulator sickness. The occurrence of simulator sickness is directly related to the aspect of adaptation. This is evident in the consideration of attempts that led to reduced symptoms of simulator sickness among the subjects over the course of the study after several trial drives. The occurrence and intensity of simulator sickness can be reduced through training or acclimatization.

The diagram in Fig. 10 provides an overview of the completed drives of the first test series. The requirement to drive freely for at least 25 minutes was achieved by 254 out of 457 participants (55.58%). 29 participants (6.35%) achieved times between 20 and 25 minutes. The remaining 179 participants

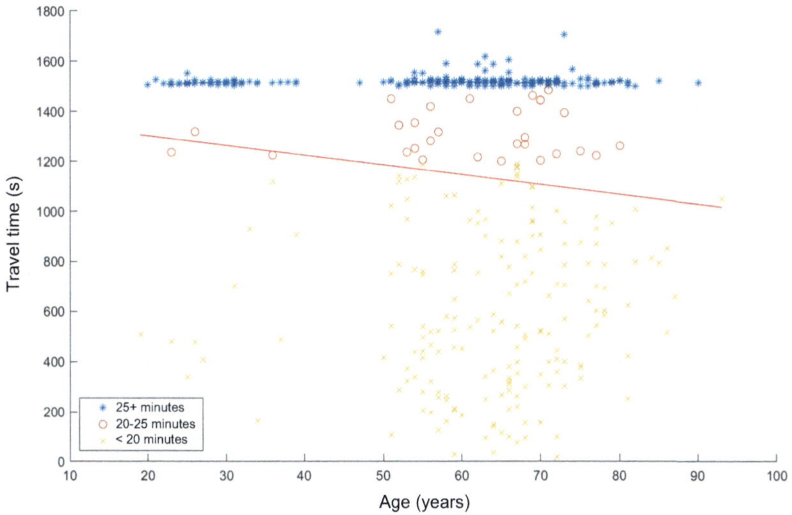

Fig. 10 Driving Time Distribution

(38.07%) terminated the experiment under 20 minutes. The diagram confirms the results from other studies that older adults are generally more susceptible to simulator sickness than younger participants (Brooks et al. 2010).

4 Driving Performance

The Index Of Performance (IOP) (Joshi, Maas, & Schramm, 2017) is calculated for an objective assessment of driving performance. The index is composed of several individual evaluations. The activity of the pedals, the driving comfort—represented by threshold values for longitudinal and lateral accelerations—and the steering activity are evaluated. The latter combines an assessment of lane keeping and steering wheel activity.

The algorithm was developed for normal traffic scenarios in simulator operation—the conditions of the first drive. However, the setup for drives two and three was changed to such an extent that a reproducible evaluation in the original form was no longer possible. This is due to the extensive highway, which did not require large steering movements and pedal work. For the highway drives, the IOP was therefore adjusted to place more emphasis on lane fidelity and steering behavior at the end of curves. This allows for a meaningful objective evaluation of driving performance in cases two and three as well.

Statements about the adaptation of the simulator can be made via the IOP (Schweig, Liebherr, Schramm, Brand, & Maas, 2018). This means that the test subjects continuously drive as in a real vehicle without, for example, completely pressing the drive and brake pedals alternately. When a person adapts the simulator, the IOP decreases. Figure 11 shows two IOP curves as examples, with an adaptation a) in which the IOP continuously decreases after about 6.11 minutes and b) without adaptation with a further increasing IOP. For a better comparison, the trend lines of the two diagrams can be used, which are drawn in after an acclimatization phase. Only the introduction of this objective evaluation of driving work allows for a somewhat reliable statement in this case as to whether the conducted investigations are adapted to reality.

Of the 463 test subjects, 102 (66 male, 36 female) adapted the system. The distribution (Fig. 12) of adaptation times over age shows almost no changes. On average, the point of adaptation shifts by only about 0.72 s/year.

Almost all participants showed a habituation effect during the third test drive with assistance systems compared to the second drive. On average, the time required for the parking tasks was reduced by 4.8 s and the total distance driven increased by 1208 m/block. An improvement in these values was observed in participants with and without changed sound settings.

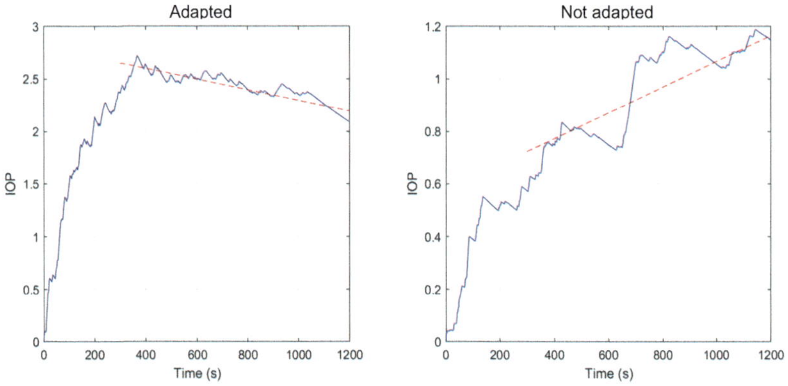

Fig. 11 **a** IOP with adaptation **b** IOP without adaptation

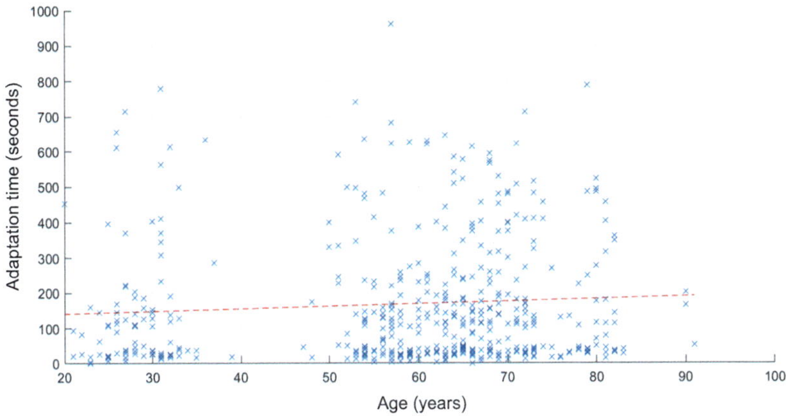

Fig. 12 Adaptation rate of the test subjects

5 Summary and Outlook

Driver assistance systems represent one of the main trends in vehicle technology alongside the electrification of the drivetrain. The aim of the work described in this book section was to lay the foundations for the development, construction and empirical verification of an age-appropriate acoustic driver assistance system. The driver assistance system to be built should be specifically tailored to the needs of the steadily growing group of individual mobile older drivers and play an economically sensible role in the product portfolio of car manufacturers and suppliers. The driver assistance system to be developed was created by integrating assistance systems that have already proven themselves in series production. To analyze and evaluate the suitability of the overall system and especially its human-machine interface, an existing driving simulator was expanded and upgraded so that the corresponding investigations could be carried out easily with a large number of test subjects.

A prerequisite for successful execution of the experiments was to provide an environment that allowed the test subjects to immerse themselves in the simulator environment in such a way that a realistic driving impression was created in each case. Added to this were the challenges expected due to the age of the test subjects. It turned out that both the ability to adapt the simulator and the timing were evenly distributed across age. Age-related effects cannot be derived from this. A significant change in driving behavior due to the changed assistance systems could also not be detected. However, a correlation could be found between the simulator sickness observed in many test subjects and their age.

For future investigations, two things are interesting for the construction and operation of simulators. Firstly, how the adaptation of the test subjects can be further improved and which countermeasures against simulator sickness can be successfully taken.

References

Brooks, J. O., Goodenough, R. R., Crisler, M. C., Klein, N. D., Alley, R. L., Koon, B. L., … Wills, R. F. (2010). Simulator sickness during driving simulation studies. *Accident Analysis & Prevention, 42*(3), 788–796.

Freund, B., & Green, T. (2006). Simulator sickness amongst older drivers with and without dementia. *Advances in Transportation Studies.*

Garcia, A., Baldwin, C., & Dworsky, M. (2010). *Gender differences in simulator sickness in fixed-versus rotating-base driving simulator.* Paper presented at the Proceedings of the Human Factors and Ergonomics Society Annual Meeting.

Joshi, S. S., Maas, N., & Schramm, D. (2017, 05–06.01.2017). *A Vehicle Dynamics Based Algorithm for Driver Evaluation* Paper presented at the 11th International Conferance on intelligent Systems and Control (ISCO), Karpagam College of Engineering Myleripalayam, Coimbatore, India.

Kennedy, R. S., Lane, N. E., Berbaum, K. S., & Lilienthal, M. G. (1993). Simulator Sickness Questionnaire: An Enhanced Method for Quantifying Simulator Sickness. *The International Journal of Aviation Psychology, 3*(3), 203–220. https://doi.org/10.1207/s15327108ijap0303_3.

Schweig, S., Liebherr, M., Schramm, D., Brand, M., & Maas, N. (2018). The Impact of Psychological and Demographic Parameters on Simulator Sickness. In *Proceedings of Simultech 2018* (pp. 91–97).

Stanney, K. M., Kingdon, K. S., & Kennedy, R. S. (2002). *Dropouts and aftereffects: examining general accessibility to virtual environment technology.* Paper presented at the Proceedings of the Human Factors and Ergonomics Society Annual Meeting.

Implementation of the Real Driving Studies—Test Vehicle and Experimental Setup in T2

Stefan Wolter, Abhinav Dhake and Carsten Starke

Contents

As part of the Alfasy project, the investigation named t2 was carried out in a real vehicle. This follows on from the simulator investigations t-1 to t1 in terms of timing. The aim of this investigation was to test the individual systems under real driving conditions. For this purpose, a test vehicle was set up with the appropriate equipment.

Dr. Stefan Wolter, Abhinav Dhake, Dr.-Ing. Carsten Starke, all Ford-Werke GmbH.

S. Wolter (✉) · A. Dhake · C. Starke
Ford-Werke GmbH, Cologne, Germany
e-mail: swolter3@ford.com

A. Dhake
e-mail: adhake1@ford.com

C. Starke
e-mail: cstarke3@ford.com

H. Proff et al. (eds.), *Age-appropriate Driver Assistance Systems*,
https://doi.org/10.1007/978-3-658-44838-7_7

1 Setup of the Test Vehicle

For the investigation in t2, Ford-Werke GmbH set up a test vehicle. It is a Ford Mondeo model year 2015 (see Fig. 1). The vehicle already has extensive special equipment. This includes a corresponding sound system with several speakers. The car also has several driver assistance systems. The sensors installed for this, especially the ultrasound system, are used for the implementation of the assistance systems. A radar system for detecting rear traffic participants was installed afterwards. A camera for lane recognition is already present.

Figure 2 shows the test vehicle schematically from a bird's eye view. The existing speakers of the standard speaker system are shown in gray. The additionally installed speakers are highlighted in black. Although the test vehicle already has a fairly high number of speakers, more were installed to ensure appropriate scalability of the system. This affects the A-pillars at the front, both right and left. Furthermore, the D-pillars at the rear, also right and left. Due to the distribution of the existing speakers marked in Fig. 2, it becomes clear that otherwise neither front nor rear directional acoustic signals could be adequately reproduced. The existing speakers along the left and right side of the vehicle cover these areas. There is also a speaker in the middle of the dashboard. As a special feature, speakers were installed in the headrests of the seats in the second row. A subwoofer installed in the trunk is also standard, but this does not play a role in this investigation due to the poor localizability of low-frequency acoustic signals.

Fig. 1 Ford Mondeo Test Vehicle

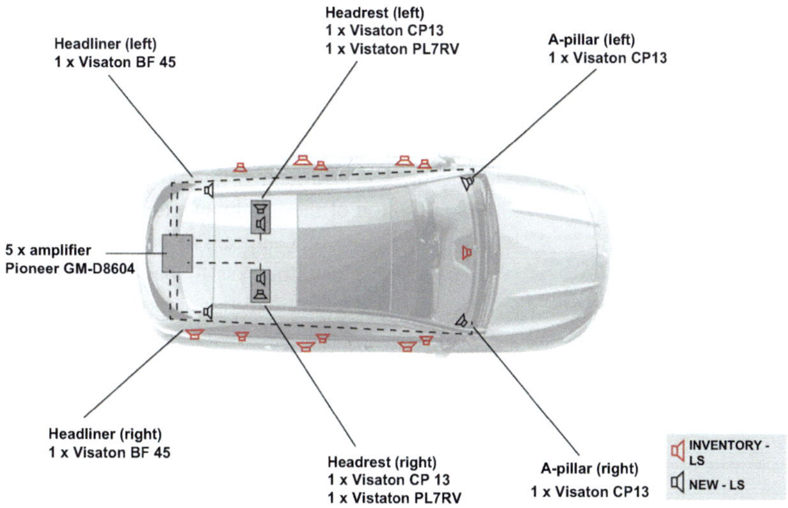

Fig. 2 Speaker Positions of the Test Vehicle

Figure 3 shows further installations of the test vehicle, primarily for controlling the speaker system. Most of the installed components are located in the trunk for space reasons. This includes a second battery to ensure an appropriate power supply, which is connected to the vehicle battery. A voltage converter ensures the supply with a voltage of 230 V. Since the vehicle network standardly has 12 V, otherwise not all additionally installed components could be supplied with the necessary operating voltage. All components in the trunk are installed on an MDF base plate. In addition to the mentioned components needed for power supply, these are the devices needed to control the audio components, such as the audio interface, which is directly connected to the amplifiers. The latter are in turn connected to the existing speakers and the additionally installed speakers. The connection is made via a terminal block. A car PC, alternatively a notebook or a dSpace MicroAutobox can be used for the control engineering implementation of the driver assistance systems. These control the speakers based on the input signal of the sensors. The software Matlab Simulink is used for this. The used computer receives its signals via the vehicle's CAN-BUS system, on which the signals of the sensors are available. Access to the CAN-BUS was also created in the trunk. The wiring for the speakers runs through the entire vehicle. An emergency stop switch for the entire power supply of the installed components is installed on the

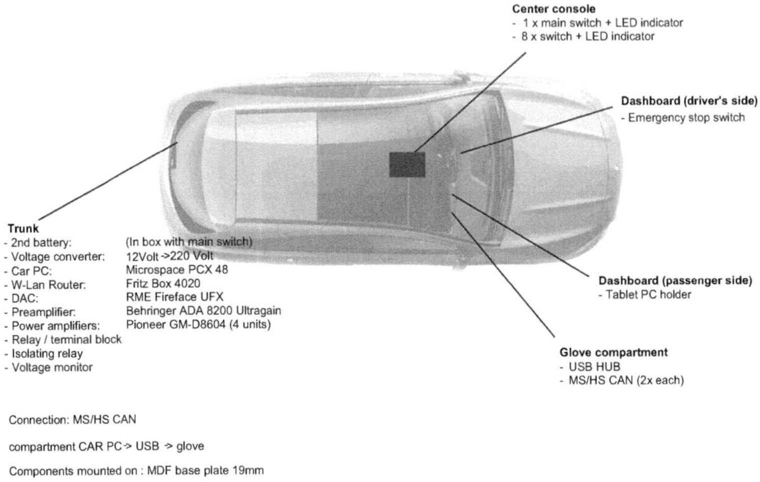

Fig. 3 Installations of the Test Vehicle

dashboard in the vehicle cockpit. In the center console there are a main switch and switches for the power supply of individual components. A further schematic representation of the components in the trunk can be found in Fig. 4. A photo of the setup is shown in Fig. 5.

A special feature concerns the installation of the speakers in the A-pillars on both sides. Since the additional speakers could not be integrated into the existing A-pillar coverings, the covering had to be redesigned. The redesign was printed on a 3D printer. Figure 6 shows the printout before installation in the test vehicle.

Figure 7 schematically shows the positions of the ultrasonic sensors of the test vehicle, the camera for lane detection, and the installed radar sensors.

2 Experimental Setup T2—Systems in the Low-Speed Range

The t2 study was divided into two parts for organizational and safety-related reasons. This chapter deals with the first part. Here, the focus was on the driver assistance systems, which operate in the low-speed or walking speed range. These are the cross-traffic assistant and the acoustic parking aid. Furthermore, the tire pressure monitoring system. The cross-traffic assistant receives the information from

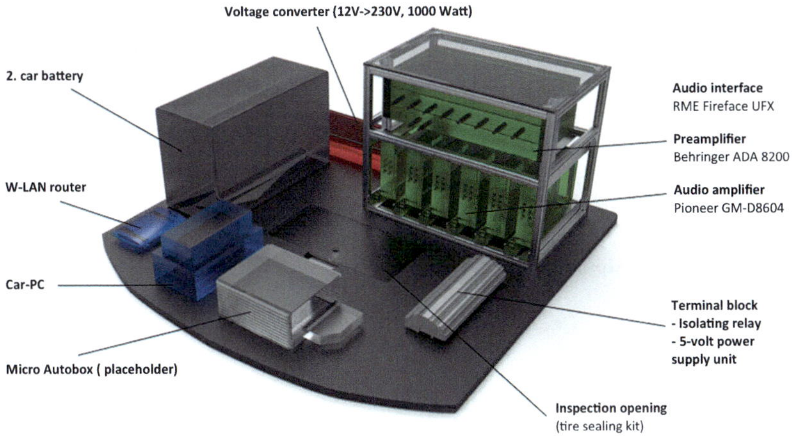

Fig. 4 Trunk Equipment of the Test Vehicle (Schematic)

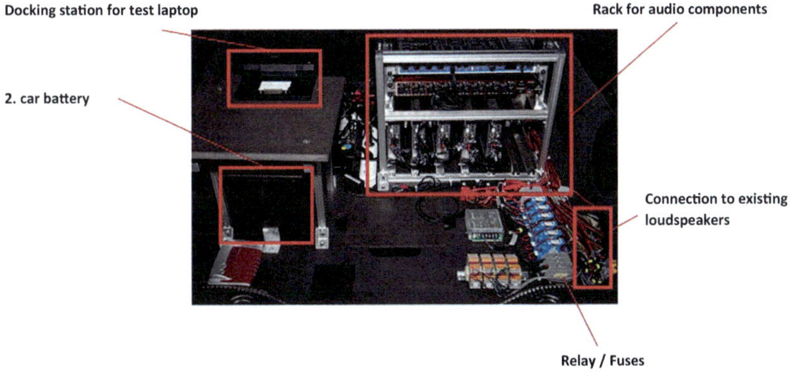

Fig. 5 Trunk Equipment of the Test Vehicle

the rear radar sensors. Figure 8 shows an example of how the system works. A vehicle comes as an obstacle from the rear left while the test vehicle is reversing out of a parking space. Then, the acoustic warning is activated. The driver is thereby prompted to brake. The system itself also works with cyclists and pedestrians.

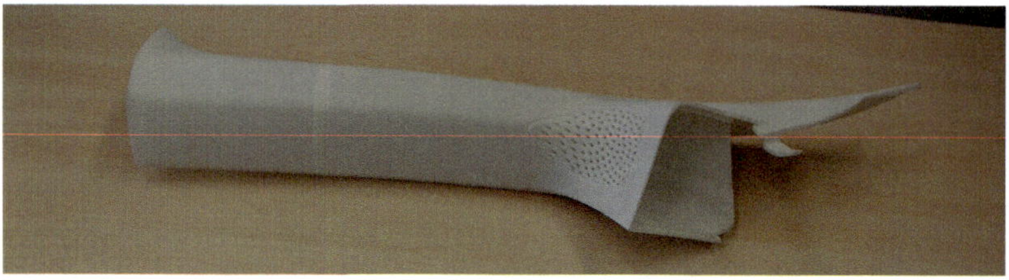

Fig. 6 3D printed A-pillar covering

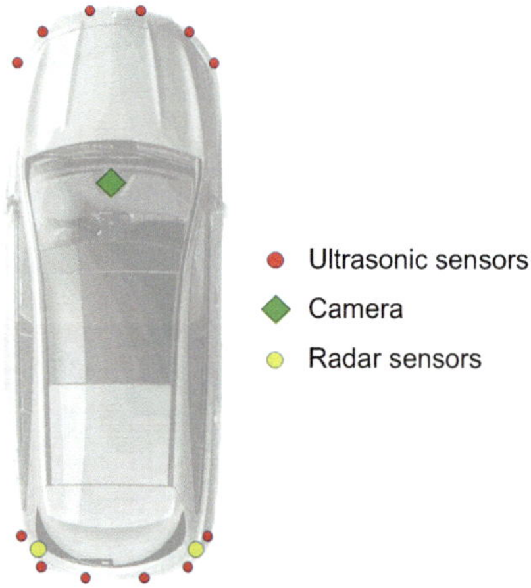

Fig. 7 Ultrasonic sensors, camera, and radar sensors of the test vehicle

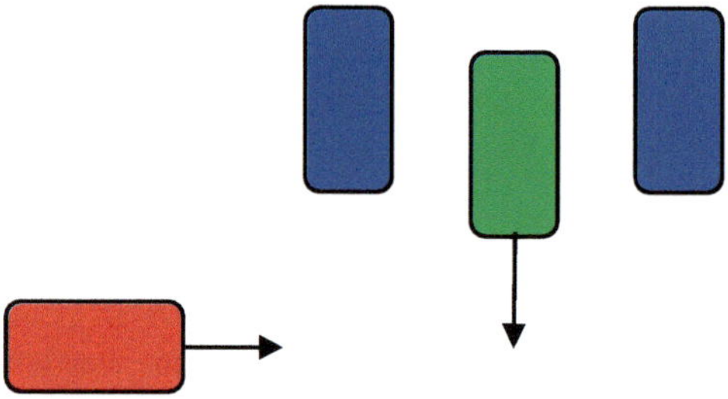

Fig. 8 Cross-traffic assistant

Fig. 9 Parallel parking

In Fig. 9, the scenario of the acoustic parking system is shown using the example of a parallel parking space. The ultrasonic sensors around the vehicle detect various obstacles. These obstacles are passed on to the driver via the acoustic system. It works the same way for perpendicular parking spaces, as well as when leaving a parking space.

The tire pressure monitoring system acoustically indicates whether a tire is losing air.

Table 1 shows the acoustic output modes of the systems under investigation. For all three systems, there is the non-directional variant of acoustic feedback. In all cases, the relevant sound was played from the speaker in the dashboard. This is the baseline variant, as the sound cannot be spatially associated with the corresponding event. The static directional variant was also implemented for all mentioned systems. In this case, the sound is played from the speaker closest to the obstacle or event. However, it still only involves one speaker. The dynamic directional variant is only applied to the cross-traffic assistant. In this case, the sound runs over several speakers. If the obstacle approaches from the rear left, the rear left speaker is activated first. The sound signal then moves from the rear left over several speakers to the rear right, thereby tracing the path of the detected obstacle. The use of a dynamic sound for the acoustic parking aid and the tire pressure loss warning does not make sense and was therefore not implemented and tested. Figure 10 schematically shows the output modes.

Figure 11 shows the test track schematically from a bird's eye view. It is a cordoned off area of the parking lot of the horse race course in Cologne. The starting position during the customer study was the fourth parking slot for each test person. There are also two parallel parking slots. Parking slot No. 1 is only limited at the

Fig. 10 Visualization of the output modes (left non-directional, right directional, static or dynamic)

Table 1 Investigated systems and output modes

	Acoustic parking aid	Cross-traffic assistant	Tire pressure loss warning
Non-directional	X	X	X
Static directional	X	X	X
Dynamic directional		X	

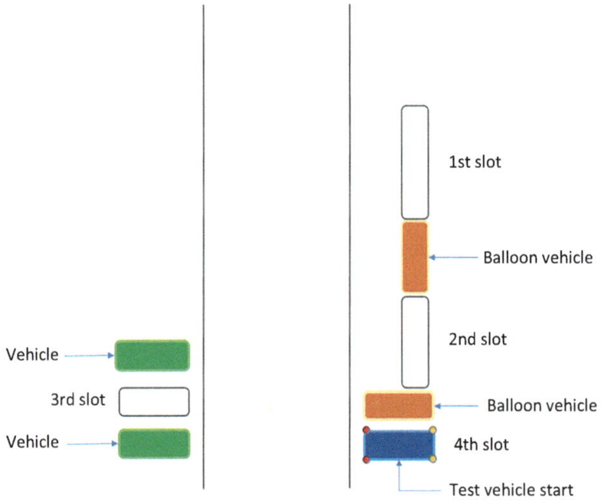

Fig. 11 Test track bird's eye view

Fig. 12 Photo of the cordoned off test track

Fig. 13 Photo of parking space No. 2 with balloon vehicles

back and is therefore a longer parking slot. Parking slot number 2 behind it is quite narrow. The length of a total of 5.70 m was limited at the front and back by inflatable balloon vehicles. The third parking slot is perpendicular and limited by two vehicles on the sides. Each run of events in the customer study ended in the fourth parking slot. The fourth and last parking slot was used for the cross-traffic assistant. Also for the event with the tire pressure monitoring system. The parking in and out of this and all other parking slots served to evaluate the acoustic parking aid.

Figure 12 shows a photo of the test track with barrier tape. Figure 13 shows the two balloon vehicles, which limit parking space number 2. Parking space number 2, with its length of 5.70 m, posed a special, albeit not unrealistic, challenge. Therefore, it was sensible to limit this parking space with the help of balloon vehicles. Figure 14 shows a photo of parking space number 3.

The participants for this study were recruited through the University of Duisburg-Essen. Details about the participants can be found in Chapter "Older Drivers: Investigation and Sample in the ALFASY Project"of this book.

Upon the arrival of a participant, they were asked about demographic data. Following this survey, a so-called localization screening was already carried out in the test vehicle. Hereby, the participant, sitting in the driver's seat of the test vehicle, was exposed to sounds from different directions. The sound used here is one of the sounds of the acoustic parking aid. For this, the speaker on the dashboard in the middle was used . Furthermore, speakers at the front left and right, as well as at the back left and right. These were played in different orders based on a randomization list. If the direction of the sound or the position of the speaker was incorrectly identified, the participant was informed about this error. After the first round of all positions, there was a second round in which these were played to the participant again with a specific information about the direction.

Then the actual study began. At the beginning, the participants were explained the driver assistance systems and the possible sound variants. The summaries used for this can be found in Table 2. The participants always started in parking space number 1. The sequence of parking spaces described above was applied for each of the conditions. These are the non-directional, the statically directional, and the dynamically directional variant. While the cross-traffic assistant has all three variants, the dynamically directional variant is omitted for the tire pressure loss warning and the acoustic parking aid. When driving through the course, the

Fig. 14 Photo of parking space No. 3

Table 2 Explanation of the systems examined

Acoustic parking aid

The parking aid uses a network of 12 ultrasonic sensors around the vehicle to warn of obstacles such as parked cars or pedestrians when parking (see image).

Parking aid for reverse parking

Cross traffic assistant

The cross-traffic assistant monitors the space to the left and right behind the vehicle when reversing out of a parking space in order to warn of crossing cars, cyclists or pedestrians (see image). The radar sensors cover a range of 20 meters on each side.

Cross traffic when parking

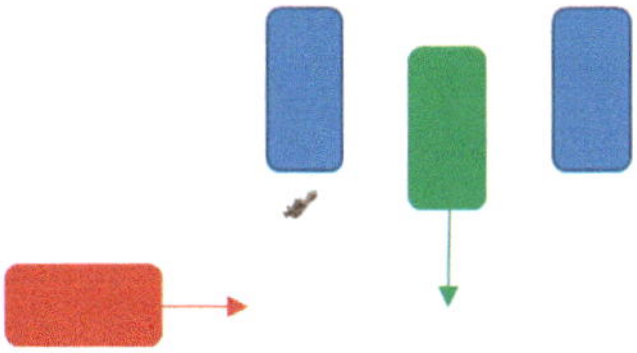

Tire pressure monitoring system

The tire pressure monitoring system constantly checks the vehicle's tire pressure. In the event that a tire suddenly loses pressure, the driver is informed immediately.

Acoustic variants

- Non-directional, "sounds only ever come from one loudspeaker, regardless of the position of the obstacle"
- Static directional, "sounds only come from the loudspeaker facing the respective obstacle"
- Dynamic directional (only for cross traffic), "sounds move across the speakers to indicate the movement of the obstacle"

variants were presented to the participants in a randomized order. However, the sequence of the individual driver assistance systems was always blocked. Starting with the acoustic parking aid, each run for an acoustic variant always ended with the event of the cross-traffic assistant, followed by the tire pressure loss warning. Since the cross-traffic assistant was the only one with three acoustic variants, there was always a single run with this at the end. The entire process for a participant lasted up to 90 minutes.

The University of Duisburg-Essen conducted its survey with a final questionnaire as part of this study. The results of this survey can be found in Chapter "Market Potentials of Older Drivers Willingness to Pay and Acceptance of Age-Appropriate Driver Assistance Systems"of this book.

3 Experimental Setup t2—Systems in the Higher Speed Range

The second part of the study t2 was separated from the first part for safety reasons and due to the higher speeds was carried out exclusively at the Ford Lommel Proving Ground in Belgium. The blind spot assistant and the lane departure warning were subjected to the test as systems with extended acoustics.

Figure 15 shows the general functioning of the blind spot assistant. The vehicle on the right side is equipped with the corresponding assistance system. Rear-facing radar sensors detect vehicles up to 15 m behind the end of the vehicle, triggering a usually visual warning. In the case of this study, an acoustic one. Since the exterior mirrors do not cover the entire area next to the vehicle (so-called blind spot), this is generally a very useful driver assistance system. A critical scenario as shown in Fig. 15, where an accident with the car on the left side is imminent, is thus avoided.

Figures 16 to 18 show the individual use cases of the study conducted in t2. Figure 16 shows how the test vehicle on the two-lane test track is overtaken by a

Fig. 15 Blind Spot Assistant Basic Principle

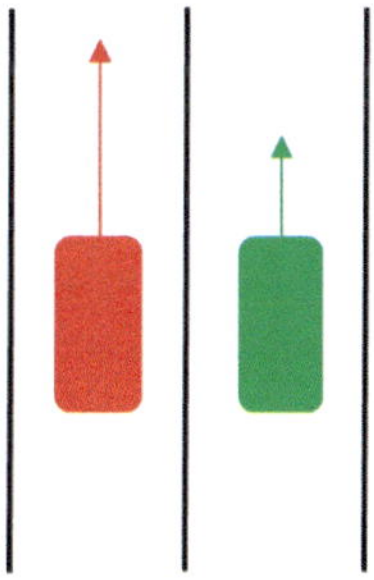

Fig. 16 Blind Spot Assistant—Use Case Being Overtaken

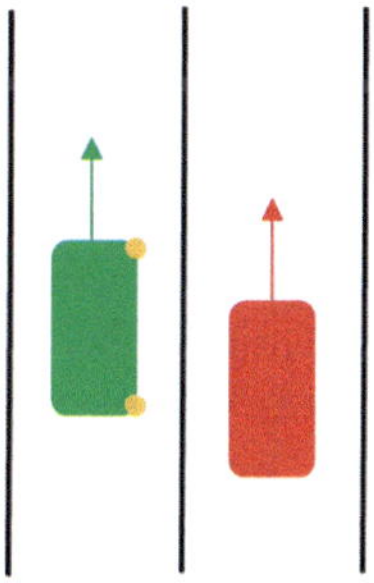

Fig. 17 Blind Spot Assistant—Use Case Overtaking

target vehicle driving on the left side. The target vehicle is detected by the system and a corresponding warning is triggered in the test vehicle.

Figure 17 shows another use case where the test vehicle overtakes the target vehicle from the left lane. There is no standard warning if a vehicle is in the blind spot. This would not be useful for many overtaking maneuvers on the left lane. Instead, the aim of this use case is to trigger another warning level. This is done by activating the turn signal. A particularly noticeable warning tone is activated to alert the driver to the danger of the overtaking process.

Figure 18 shows a use case that is analogous to the one in Fig. 17. The test vehicle is driving on the right lane and is being followed by another vehicle, which remains in the blind spot. The driver again activates the turn signal, triggering a particularly noticeable warning tone.

The lane departure warning works by detecting the lane marking through a camera system. If the lane is unintentionally left, i.e., without activating the turn signal, a warning is given depending on the specific implementation, usually by the vibration of the steering wheel. In relation to the lane departure warning,

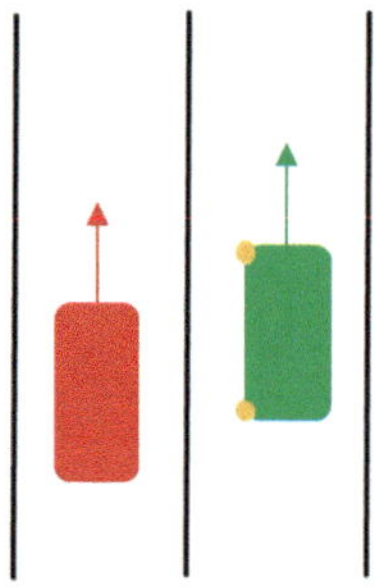

Fig. 18 Blind Spot Assistant—Use Case Being Overtaken with Attempt to Overtake

Fig. 19 shows the use case applied. The test car is driving on the right lane of the two-lane test track. To trigger the acoustic warning of the lane departure warning system, the driver moves to the left over the lane marking. This is done without activating the turn signal. The driver then moves back to the right lane to start the next run from there. This run was always covered by another vehicle from behind to ensure safety on the test track.

As with the first part of t2, there were again three different implementations of the acoustics for both the lane departure warning and the blind spot assistant. On the one hand, the warnings were presented non-directionally, i.e., solely via the speaker in the middle of the vehicle's dashboard. In the statically directional condition, the position of the activated speaker depends on the position of the obstacle or the lane being crossed. In the dynamically directional variant, there is not only a dependency on the position. The sound also moves along the obstacle in case of the blind spot assistant. In case of the lane departure warning, it moves along the lane that is currently being left. With the blind spot assistant, the sound runs on the respective side from right or left rear to front. In case of lane departure

Fig. 19 Lane Departure Warning—Crossing the Marking

Table 3 Examined driver assistance systems and output modalities

	Lane departure warning	Blind spot assistant
Non-directional	X	X
Static directional	X	X
Dynamic directional	X	X

warning, on the respective side from the middle to the front. Table 3 shows the experimental conditions for the two driver assistance systems examined.

As part of the actual study, a section of the Ford Lommel Proving Ground in Belgium was selected for the study implementation. This is an oval-shaped, 6 km long track with two lanes. Per round, this meant a driving time of about 4 min. For the study, 20 people were selected, all of whom are Ford employees. Due to appropriate safety precautions, only participants with a minimum classification as test drivers were admitted. Furthermore, there was a detailed examination of the safety standards by the responsible bodies on site.

The study implementation began with a collection of demographic data. This was followed by a localization screening in the form that sounds were played from five different directions with the vehicle stationary. The origin of these sounds had to be recognized by the participants. If the direction of a sound was incorrectly identified, the participant was informed and educated about the actual direction. As with the first part of the study with the driver assistance systems in the area of low speeds, the same speakers were activated in a randomized order. After this round, all sounds were played again with an explanation of the direction. The two systems to be evaluated, lane departure warning and blind spot assistant, were explained to the participants before the actual drive, as well as the different acoustic variants to be tested (i.e., non-directional, static directional, and dynamic directional). Table 4 shows the explanation of the systems used. The explanation of the acoustic variants was analogous to the description in Table 2. The actual drive began with an introductory round on the test track. The participants were supposed to maintain a speed of 90 km/h. Based on the randomized order, the participants either started with the lane departure warning or with the blind spot assistant. Within the individual systems, the assignment of the acoustic variants was also randomized, i.e., the sequence of the non-directional, static directional, and dynamic directional variants. While only one use case with the departure of the lane was examined for the lane departure warning, the experience of the blind spot assistant was divided into three different use cases: As already

Table 4 Explanation of the systems examined

The blind spot assistant uses radar sensors to detect objects in the blind spot to the left and right of the vehicle. This primarily involves other cars, although bicycles, for example, can also be detected. The system goes beyond the blind spot area and detects objects up to 15 meters behind the end of your own vehicle, e.g. to assist with lane changes on the freeway (see image). In this case, the warning is given acoustically. If a vehicle is in the blind spot and the blinker is active, another sound is played. 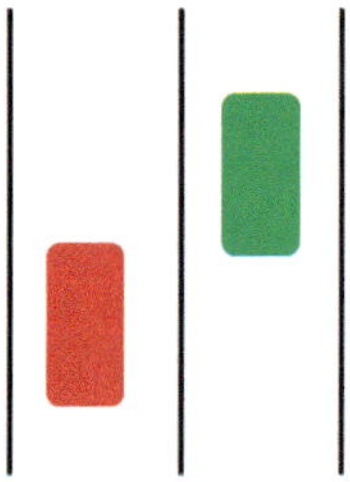 Vehicle in blind spot (right: own vehicle; left: vehicle in blind spot)
The lane departure warning alerts the driver via a warning system if the vehicle drifts out of lane (see image). The lanes are detected by a camera system. The lane departure warning system is switched off briefly when the blinker is activated so as not to trigger a false alarm when changing lanes. The warning is given acoustically and via the vibration of the steering wheel. 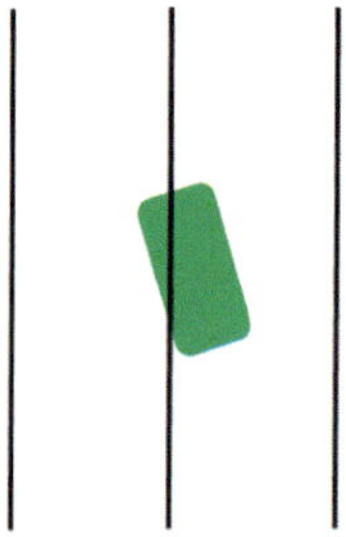 Vehicle leaves lane, lane departure warning is activated

described above, overtaking, driving to the right of the target vehicle with activated blinker, and finally driving on the left side, also with activated blinker. For organizational reasons, the sequence of use cases was always presented in this order. Trying out each acoustic variant per use case lasted about half a round on the oval test track. This resulted in at least three activations each time. In the last two use cases of the blind spot assistant, only the non-directional and static directional variant were presented, as this is an escalation level. After each round, there was a short post-survey on acceptance and the question of how helpful the system is. In the final post-survey, it was recorded which acoustic variant is preferred for the lane departure warning and for the blind spot assistant. Furthermore, the van der Laan scale (van der Laan, Heino & de Waard, 1997) was used for the evaluation of the different acoustic variants across both driver assistance systems. The entire study duration lasted one hour.

4 Results T2—Systems in the Area of Higher Speed

In case of the second examination of t2, as described, several scales were applied to evaluate the driver assistance systems. The most important results are presented below.

Figure 20 shows the evaluation of the blind spot assistant. The three acoustic variants were compared on a 5-point scale from 1= "strongly disagree" to 5= "strongly agree". In each case, it was asked whether the system that plays non-directional, statically directional, or dynamically directional sounds is preferred. The statically directional variant is rated best, followed by the dynamically

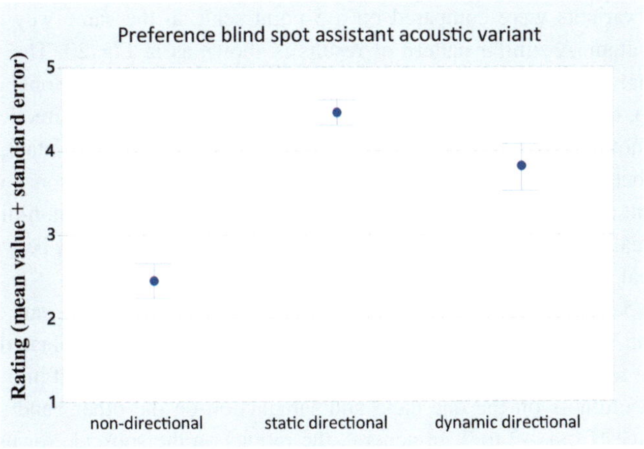

Fig. 20 Evaluation of the acoustic variants of the blind spot assistant

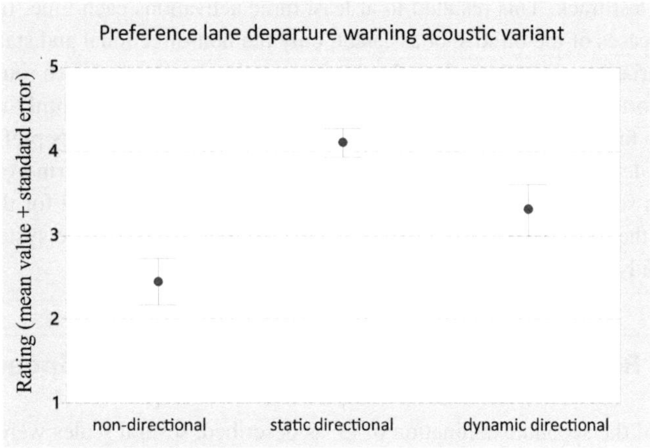

Fig. 21 Evaluation of the acoustic variants of lane departure warning

directional and, with a large gap, the non-directional acoustic variant. The difference between the statically directional and dynamically directional variant is not significant in the t-test. The difference between the other two paired comparisons is significant, with $p<.001$ between non-directional and statically directional and $p<.01$ between non-directional and dynamically directional.

Figure 21 shows the evaluation of lane departure warning. Again, the three acoustic variants were compared on a 5-point scale in the same way as in the previous item. A similar pattern of results is shown as in Fig. 20. The statically directional variant is rated best, followed by the dynamically directional and, by a large gap, the non-directional acoustic variant. However, the two directional variants are not rated quite as positively as in case of the blind spot assistant. The difference between the statically directional and dynamically directional variant is significant at the $p<.05$ level in the t-test. The difference between non-directional and statically directional is significant at the $p<.001$ level and that between non-directional and dynamically directional at the $p<.05$ level.

Table 5 shows the evaluation of the acoustic variants on the van der Laan scale (van der Laan et al., 1997). Here, these were generally evaluated without reference to the specific systems. The van der Laan scale is divided into two subscales, usefulness on the one hand and satisfaction on the other hand side. The scale ranges from −2 to 2. In general, the ratings on the subscale for usefulness are higher than on the one for satisfaction, but the pattern of the sequence is the

Table 5 Evaluation of the acoustic variants on the van der Laan scale

Mean (Standard error)	Usefulness	Satisfaction
Non-directional	−0.01 *(0.12)*	−0.35 *(0.10)*
Statically directional	1.24 *(0.07)*	0.79 *(0.08)*
Dynamically directional	0.75 *(0.11)*	0.33 *(0.13)*

same. The evaluation of the statically directional acoustic variant is the best, followed by the dynamically directional and finally the non-directional. The difference between the non-directional and the statically directional is significant at the $p < .001$ level on both scales. All other paired comparisons are significant at the $p < .05$ level.

Overall, both the blind spot assistant and the lane departure warning show a preference for the static directional acoustic variant. The dynamic directional falls behind, and the non-directional variant performs worst. This result can be explained by the fact that the non-directional variant provides the least information value. In the case of the two directional variants, the encoding of the location, i.e., the obstacle or the departed lane, is included. When comparing the two directional variants, it should be noted that the static directional variant already contains enough information about the location in both cases. The additional dynamics of the sound do not provide any further advantage and may even be perceived as disturbing or irritating. Based on these data, the implementation of a static directional acoustic feedback for the mentioned driver assistance systems represents the best variant.

References

Van der Laan, J.D., Heino, A., & de Waard, D. (1997). A simple procedure for the assessment of acceptance of advanced transport telematics. *Transportation Research – Part C: Emerging Technologies*, *5*, 1–10.

Neuropsychological and -Physiological Correlates of the Driving Behavior of Older Drivers Within Simulated Environments

Magnus Liebherr, Melanie Zerr and Matthias Brand

Contents

Primary mobility represents one of the most important factors in maintaining the quality of life and independence of older people. Due to the increasing complexity of road traffic, increasingly higher demands are placed on the attention, concentration, and information processing of all road users. The increasing demands on cognitive abilities are countered by an age-related decline in cognitive performance. Increased accident rates and a deterioration in driving performance with

Dr. Magnus Liebherr, Melanie Zerr, B.Sc., Prof. Dr. Matthias Brand, all University of Duisburg-Essen

M. Liebherr (✉) · M. Zerr (✉) · M. Brand
Universität Duisburg-Essen, Essen, Germany
e-mail: magnus.liebherr@uni-due.de

M. Zerr
e-mail: melanie.zerr@stud.uni-due.de

M. Brand
e-mail: matthias.brand@uni-due.de

H. Proff et al. (eds.), *Age-appropriate Driver Assistance Systems*,
https://doi.org/10.1007/978-3-658-44838-7_8

increasing age are also reported (e.g., Wood, 2002). Driving simulators provide a good opportunity to investigate potential influence and interaction effects on driving performance. While field studies are often characterized by increased complexity and the occurrence of various difficulties that reduce the comparability of driving conditions, simulators allow relevant factors to be investigated at a highly standardized, but also cost- and time-efficient level (Desmond & Matthews, 1997). In addition, the risk of accidents or injuries that can occur in real situations can be eliminated by using simulated environments. While simulators were originally developed in the context of flying, they are now most commonly used in driving behavior studies (Weinberg & Harsham, 2009). In contrast to early simulators, which were very simple and rudimentary, current simulators are highly developed to reproduce as realistic behaviors as possible (Bayarri, Fernandez, & Perez, 1996). Although previous studies could not find a difference between driving behavior in simulators and on real roads (Underwood, Crundall, & Chapman, 2011), some authors are still concerned about the validity of the findings collected in driving simulators. Kemeny and Panerai (2003) found, for example, that driving simulators do not provide all relevant visual cues for drivers to feel and behave comparably to a real driving situation. In addition, Owsley and McGwin Jr (2010) point to a coarse visual representation in driving simulators, which can lead to invalid measurements. In addition, studies show that participants tend to drive slower in the simulator, which indicates a probable uncertainty regarding the virtual reality shown in the simulator (Godley, Triggs, & Fildes, 2002; Klee, Bauer, Radwan, & Al-Deek, 1999). Due to the existing differences between real and simulated driving, problems such as simulator sickness (for a detailed explanation, see next section) and a lack of adaptation to the simulator often occur when using simulators (Brooks et al., 2010), which in turn can affect the driving performance of the participants. The process of aging is often discussed in this context as a factor that could influence the occurrence of simulator sickness, the process of adaptation to the simulator, and simulator-associated stress reactions. Evidence from previous simulator studies points to a reduced driving performance (Andrews & Westerman, 2012), an increased risk of accidents (Lee, Cameron, & Lee, 2003) as well as increased stress (Cantin, Lavallière, Simoneau, & Teasdale, 2009) in older compared to younger drivers.

This chapter addresses these aspects based on three investigations. It describes results on the topics 1) Cognitive abilities, simulator adaptation, and simulator sickness, 2) Stress reactions when driving in the simulator, and 3) Attention and driving behavior, each taking into account the age of the drivers.

1 Cognitive Abilities, Simulator Adaptation, and Simulator Sickness

Sahami, Jenkins, and Sayed (2009) describe simulator adaptation as the point at which existing driving skills are successfully transferred to the simulator. To ensure this, predefined practice times (Horberry, Anderson, & Regan, 2006), practice routes (Lewis-Evans & Charlton, 2006), and queries regarding the well-being of the participants (Takayama & Nass, 2008) were used in previous studies. However, these approaches are not methodologically sufficient as they do not take into account the individual abilities of the respective participants. An objective measure of the point of simulator adaptation is described by McGehee, Lee, Rizzo, Dawson, and Bateman (2004). Based on the measurement of steering movements, the authors identified an average adaptation to the simulator within about 3 minutes after the start of the simulation. In addition, they reported no significant differences between younger and older participants.

Simulator sickness is a phenomenon occasionally reported in connection with simulated environments and includes symptoms such as headaches, (cold) sweating, dry mouth, dizziness, disorientation, drowsiness, nausea, vertigo, and vomiting (Balk, Bertola, & Inman, 2013; Brooks et al., 2010). Previous studies describe dropout rates due to symptoms of simulator sickness from 5% to 30% (Cobb, Nichols, Ramsey, & Wilson, 1999; Stanney, Kingdon, & Kennedy, 2002; Stanney, Mourant, & Kennedy, 1998). Some studies identified gender-specific differences with a lower dropout rate in men (Garcia, Baldwin, & Dworsky, 2010). In addition, earlier results show an increase in symptoms of simulator sickness in older adults compared to younger ones (Brooks et al., 2010), which is explained by reduced experience with simulated environments (Domeyer, Cassavaugh, & Backs, 2013). Regarding the influence of cognitive functions, on the one hand, no effects on the occurrence of symptoms of simulator sickness could be identified (Mullen, Weaver, Riendeau, Morrison, & Bédard, 2010). On the other hand, studies report an increased rate in people with reduced visual-spatial performance (Kawano et al., 2012). Studies with cognitively impaired participants describe a relationship between cognitive functions and the incidence rate of simulator sickness (Freund & Green, 2006). Based on existing theories of simulator sickness, a connection between adaptation to a virtual environment and the occurrence of symptoms of simulator sickness can be assumed (Claremont, 1931; Ebenholtz, 1992; Money & Myles, 1974; Reason, 1978). This is also reflected in empirical data showing that better adaptation to new virtual environments in the simulator reduces the occurrence of symptoms (Mackrous, Lavallière, & Teasdale, 2014).

Previous studies describing reduced cognitive abilities with increasing age (Wecker, Kramer, Hallam, & Delis, 2005), together with findings that older people have low adaptation to new technologies (Roupa et al., 2010), suggest that older individuals may have difficulty adapting to virtual driving environments and are more susceptible to symptoms of simulator sickness. Furthermore, it is assumed that individuals who have adapted to the system in previous use of simulators differ in the occurrence of symptoms of simulator sickness from those who have not adapted. To gain a better understanding of individual factors such as age, logical thinking, cognitive flexibility, and attention/concentration in the context of simulator adaptation and simulator sickness, these aspects were addressed within the present project.

For this, 414 participants (M = 61.69 years, SD = 12.66, 25-89 years, 153 women) of the total sample described in Chap. 2 performed different test procedures for attention [d2-test] (Brickenkamp, 1962), cognitive flexibility [Trail Making Test-B] (Reitan, 1971) and logical thinking [Performance Test System Subtest 4] (Horn, 1983). Within the *d2-test*, the participants are supposed to cross out all "d" characters on a sheet of paper, which are combined with two strokes either above or below the character (Brickenkamp, 1962). On the one-sided sheet of paper, there are 14 rows, each row with 47 "p" and "d" characters, which are combined with one to four strokes above and/or below each letter. Participants have 20 seconds per line. The test performance is quantified by the sum of the incorrectly crossed characters and the missing correct characters. As a result, a higher value reflects a poorer performance. The *Trail Making Test* consists of two parts (A & B) in which the participants connect numbers (A) or numbers and letters (B) on a sheet of paper (Lezak, 1995). Within part B, 12 letters (A-K) and 13 numbers (1–13) must be connected as quickly as possible alternately. The result variable is represented by the time the participants need to complete the task. If errors are made, the participants are immediately made aware of this by the experimenter, which is reflected in an increase in the required time. The *LPS-4* is a subtest of the performance test system and represents the logical thinking of the test persons (Horn, 1983). The test includes 40 lines, each with 9 numbers and/or letters. Each line has a logical sequence with a letter or a number that does not fit into this sequence. The aim of the task is to identify and mark the non-matching character. The level of difficulty increases over the 40 lines. The available time is limited to eight minutes. The result parameter is the sum of the correct lines. Accordingly, a higher value in the test means a better result. The *time of adaptation* was determined based on the "Index of Performance" (Joshi, Maas, & Schramm, 2017) of the first drive in the simulator. *Simulator sickness* was quantified by the time of termination due to the occurrence of corresponding symptoms.

Table 1 Correlation of age, time in simulator and cognitive abilities

	2	3	4	5
1. Age	–,103*	,351**	,475**	–,387**
2. Time in simulator	–	–,054	–,069	,052
3. d2		–	,470**	–,429**
4. TMT(B)			–	–,522**
5. LPS-4				–

$*p \leq ,050$. $**p \leq ,010$

The results show a significant correlation between age and time in the driving simulator (M = 1,018.18 s, SD= 454.94), as well as the tested cognitive abilities of attention (M= 158.81 errors, SD = 40.15), cognitive flexibility (M = 83.29 s, SD = 34.33) and logical thinking (M = 25.19 count, SD= 4.14) (see Table 1).

Summarizing the results, it can be stated that increasing age is associated with poorer attention performance (d2-test), decreased cognitive flexibility (TMT-B), reduced logical thinking performance (LPS-4), and a shorter time spent in the simulator (increasing age led to individuals having to quit prematurely due to symptoms of simulator sickness). Overall, 193 participants (46.62%) had to quit prematurely due to symptoms of simulator sickness (e.g., headaches, cold sweats, dizziness, nausea, and vomiting). Within the sample, 90 participants (M = 60.36 years, SD = 14.39) adapted to the driving simulator. The average adaptation time was 549.97 s (SD = 280.68). No correlations could be identified between age and simulator adaptation time. However, a correlation was identified between the performance in the d2 test and the simulator adaptation time. The results of the 414 participants regarding the correlations between age and cognitive abilities could be replicated (see Table 2).

Moderated regression analyses with age as a predictor and the moderators a) intelligence, b) cognitive flexibility, c) attention, as well as the interaction of age and cognitive functions on the intensity of simulator sickness symptoms (criterion) were calculated. In the first step, age significantly explained the variance of simulator sickness, $\Delta R2$ = ,011, $\Delta F(1, 412)$ = 4,455, p = ,035. In the second step, intelligence (ΔR^2 = ,011, $\Delta F(1, 411)$ = ,067, p = ,796), cognitive flexibility (ΔR^2 = ,011, $\Delta F(1, 411)$ = ,222. P = ,638) and attention (ΔR^2 = ,011, $\Delta F(1, 411)$ = ,146, p = ,702) did not lead to a significant increase in variance explanation. In the third step, the interaction between age and intelligence (ΔR^2 = ,014, $\Delta F(1, 410)$ = 1,397, p = ,238), age and cognitive flexibility

Table 2 Correlation of age, simulator adaptation, and cognitive abilities in individuals who have adapted to the simulator

	2	3	4	5
1. Age	−,150	,315**	,461**	−,364**
2. Time in Simulator	–	,222	−,051	,009
3. d2		–	,401**	−,352**
4. TMT(B)			–	−,599**
5. LPS-4				–

$N = 90$; $^*p \leq ,050$. $^{**}p \leq ,010$

($\Delta R^2 = ,013$, $\Delta F(1, 410) = ,816$, $p = ,367$) and age and attention ($\Delta R^2 = ,013$, $\Delta F(1,410) = ,678$, $p = ,411$) were also not significant.

Regarding the explanation of the timing of adaptation, moderated regression analyses were also calculated. Included in the calculation were age (predictor) and a) intelligence, b) cognitive flexibility, c) attention as moderators, as well as the interactions between age and cognitive functions on the timing of adaptation (criterion). In the first step, age did not significantly explain the variance of simulator adaptation $\Delta R2 = ,022$, $\Delta F(1, 88) = 2,013$, $p = ,160$. In the second step, intelligence ($\Delta R2 = ,025$, $\Delta F(1, 87) = ,210$, $p = ,648$) and cognitive flexibility ($\Delta R^2 = ,023$, $\Delta F(1, 87) = ,035$, $p = ,853$) did not lead to a significant increase in variance, but attention did ($\Delta R2 = ,103$, $\Delta F(1, 87) = 7,793$, $p = ,006$). In the third step, the interaction of age and intelligence ($\Delta R^2 = ,025$, $\Delta F(1,86) < ,000$, $p = ,990$), age and cognitive flexibility ($\Delta R^2 = ,025$, $\Delta F(1,86) = ,160$, $p = ,690$), as well as age and attention ($\Delta R^2 = ,105$, $\Delta F(1,86) = ,265$, $p = ,608$) were also not significant. Additionally, it was shown that individuals who adapted to the simulator ($M = 1395,12$, $SD = 287,90$) spent significantly more time in the simulator compared to participants who did not adapt ($M = 1053,16$, $SD = 512,17$) ($t(259,1) = -8,220$, $p < 0,001$, $d = ,823$).

2 Use of Driving Simulators and Associated Stress Reactions

For the demands that driving places on the individual, the prefrontal cortex (Jäncke, Brunner, & Esslen, 2008) and the basal ganglia are particularly relevant at the brain level (Uchiyama, Ebe, Kozato, Okada, & Sadato, 2003). Both are brain regions whose functioning can be influenced by a stress-induced increase in catecholamines and cortisol (Cabib & Puglisi-Allegra, 2012; Cohen, Braver, & Brown, 2002; Ramos & Arnsten, 2007). To quantify the acute psychophysiological effects of driving on the drivers, Yamaguchi and Sakakima (2007) measured the enzyme alpha-amylase in saliva at several measurement points (before a drive in the simulator and every three minutes during the 21-minute drive). The results show a significant increase in stress levels (operationalized via noradrenergic activity, indicated by the alpha-amylase concentrations) during the course of the drive. In addition to assessing changes in endocrine activity, heart rate is often used as a biomarker for investigating stress levels (Vrijkotte, Van Doornen, & De Geus, 2000).

In comparison of simulated and real driving, early studies show no differences in terms of stress levels (Dorn & Matthews, 1995; Gulian, Matthews, Glendon, Davies, & Debney, 1989). In contrast, Engström, Johansson, and Östlund (2005) identified differences in the physiological parameters of real driving compared to simulated driving, with real driving eliciting stronger stress responses. Johnson et al. (2011) confirmed these results. The authors report significantly higher average heart rates in road traffic compared to simulated driving.

In addition to technical aspects of the currently used simulator systems, however, some other factors must be considered in the present context. For example, previous experiences with symptoms of simulator sickness—as discussed in more detail in the previous section—must be associated with an increased stress level in future studies in the simulator. In this context, Eversmann et al. (1978) report on a stimulation of hormone secretion induced by various degrees of motion sickness. In addition, the authors identified the secretion of antidiuretic hormones as the most sensitive indicator of stress induced by motion sickness. Furthermore, studies show a connection between simulator sickness and the glucocorticoid/sympathicoadrenergic system (Otto, Riepl, Klosterhalfen, & Enck, 2006) as well as the endocannabinoid system (Choukèr et al., 2010). Stress is also induced by the expectation of a stressful event. Previous studies show that the simple instruction to give a public speech leads to increased anxiety and negative mood (Al'Absi et al., 1997) as well as an increased cortisol level (Dickerson & Kemeny, 2004; Juster, Perna, Marin, Sindi, & Lupien, 2012). Based on these findings, it can be assumed that previous experiences with symptoms of simulator sickness influence the stress level before reusing the simulator. As described in detail previously, the (non-)adaptation to simulators can be associated with the occurrence of simulator sickness. Here we were able to show that individuals who had not become accustomed to the simulator had a higher dropout rate due to simulator sickness. It can therefore be assumed that a lack of adaptation to simulated environments can lead to stress. This is in line with the model of allostatic load introduced by Sterling and Eyer (1988), according to which stress mainly arises from mental and physical adaptation to new and changing conditions (Sterling, 2012). In these situations, allostatic describes the process by which the body maintains its stability through behavioral changes (McEwen, 1998, 2010). Applied to the area of simulated environments, it can be assumed that these situations represent novel and changing conditions—especially for individuals with little or no experience with simulated environments—and therefore lead to an increased stress level. In addition, Domeyer et al. (2013) stated that initial acclimatization to the driving simulator reduces the symptoms of simulator sickness. Therefore, it can be assumed that individuals who have previously adapted to a simulator have a lower stress level before subsequent use of the simulator than individuals who have not previously adapted. We assume that the latter have a higher anticipatory stress level before a second simulator drive.

Age-related effects are often discussed in the present context. For example, Matthews, Joyner, and Newman (1999) describe a higher stress level in older adults than in younger individuals while driving in a simulator. Reimer, Mehler, Pohlmeyer, Coughlin, and Dusek (2006) report that older drivers show a higher strain compared to younger drivers during a simulator driving task. A possible explanation comes from Aldwin (1991). The author argues that older adults have less control over their environment than younger adults and therefore show a higher stress level. Further studies report increased symptoms of simulator sickness (Classen, Bewernitz, & Shechtman, 2011) and a decreased adaptation rate (Domeyer et al., 2013) in older compared to younger adults. According to the model of allostatic load, an increased stress level results from an increased strain regarding adaptation to a new environment or changing conditions. This could be another explanation for age-related differences in simulator tasks (Findlater, Froehlich, Fattal, Wobbrock, & Dastyar, 2013; Kang & Yoon, 2008). Based on previous findings, the following hypothetical model was established, which was the subject of the present investigation (see Fig. 1):

In addition to the statistical testing of the model, the present sub-study aimed to investigate the effect of adaptation to a simulator on the stress level before further use.

For the study, 164 participants ($M = 61.62$ years, $SD = 12.66$, 25–89 years, 42 women) of the total sample were included. To capture the intensity of the symptoms of simulator sickness, participants were asked to fill out a modified version of the "Simulator Sickness Questionnaire" (SSQ) online 5 days after completing the first simulator drive (Brooks et al., 2010). Prior to and immediately after the second drive in the simulator, the stress level was measured using salivary alpha-amylase.

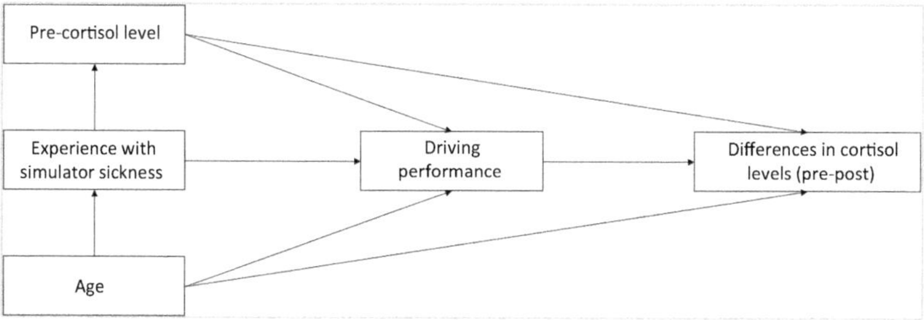

Fig. 1 Hypothetical model of stress, experience with simulator sickness, and age

Table 3 Bivariate correlations between age, experience with simulator sickness, pre-cortisol level, difference in cortisol level (before and after driving) and driving performance

	M(SD)	1.	2.	3.	4.
1. Age	61.62(12.66)				
2. Simulator Sickness Symptoms (SSQ)	3.31(2.65)	–,125			
3. Pre-Cortisol	17.25(13.48)	,073	–,064		
4. Difference in Cortisol Level (post-pre)	,63(17.48)	,188*	,244**	–,434**	
5. Driving Performance (2nd Drive)	–,03(.23)	,216**	,049	,042	,114

$^*p \leq ,050$. $^{**}p \leq ,010$

The average values of the intensity of simulator sickness symptoms (as a measure of experience with simulator sickness), pre-cortisol levels, the difference in post-pre cortisol levels, as well as driving performance are listed in Table 3.

The driving performance in the second simulator drive correlates with age, but not with experience with simulator sickness or the pre-stress level. Furthermore, the experience with simulator sickness correlates exclusively with the difference in cortisol level. As a result, the original hypothetical model could not be tested.

Based on the theoretical background of the original model, two hierarchical moderated regression analyses with age (predictor), a) driving performance, b) experience with simulator sickness (moderator) and the interaction on the difference in cortisol level (post-pre) were conducted (all variables were centralized) (Cohen, Cohen, West, & Aiken, 2003).

In the first step, age significantly explained the difference in cortisol levels, $R^2 = 0.029$, $F(1,162) = 5.914$, $p = 0.016$. Neither the addition of mileage in the second step ($\Delta R^2 = 0.006$, $\Delta F(1,161) = 0.954$, $p = 0.330$) nor the interaction of age and mileage in the third step ($\Delta R2 = 0.003$, $\Delta F(1,160) = 0.474$, $p = 0.492$) led to a significant increase in the explanation of variance. The overall model explains 2.6% of the variance in the change in cortisol levels ($F = 2.439$, $p = 0.492$). The second analysis, including the experience with simulator sickness, yielded a significant additional explanation of the variance (beyond the effect of age) in relation to the difference in cortisol levels of nearly 10% ($\Delta R^2 = 0.073$, $\Delta F(1,161) = 13.097$, $p < 0.001$). In the third step, the interaction between age and the experience with simulator sickness was not significant ($\Delta R^2 = 0.014$, $\Delta F(1\ 160) = 2.470$, $p = 0.118$). The overall model was significant and explains 10.5% variance in the change in cortisol levels ($F = 7.366$, $p < 0.001$).

Within the second simulator ride, the average pre-cortisol level of the participants who had adapted to the simulator during the first ride was 16.55 ($SD = 11.70$) and 17.67 ($SD = 14.39$) for those who had not adapted. No statistically significant group differences could be identified $t(160) = 0.493$, $p = .623$.

3 Attention and Driving Behavior

Numerous previous studies underline the relevance of attention processes in the context of driving performance and accident rate (Horberry, Anderson, Regan, Triggs, & Brown, 2006). Klauer, Dingus, Neale, Sudweeks, and Ramsey (2006), for example, identified an accident rate due to inattention of 80%. Particularly long glances inside the vehicle are often associated with accidents (Horrey & Wickens, 2007; Klauer et al., 2006; Wikman, Nieminen, & Summala, 1998). The extent of this problem increases with the growing popularity of vehicle technologies (Lerner & Boyd, 2005).

In the past, it has been shown that various attention domains are relevant for successful driving. For example, selective attention is often discussed as an important skill and limitations of this are associated with an increased accident risk and a reduced driving performance (De Raedt & Ponjaert-Kristoffersen, 2000; Lundqvist, Gerdle, & Rönnberg, 2000; Richardson & Marottoli, 2003). In addition, previous studies report relationships between visual selective attention performance (Baldock, Mathias, McLean, & Berndt, 2007) and auditory selective attention performance (Arthur Jr & Doverspike, 1992) and road driving performance and accident rate.

With the increasing popularity of phoning while driving, processes of divided attention have increasingly come to the fore in driving studies (McCartt, Hellinga,

& Bratiman, 2006; Strayer & Drews, 2007). Strayer and Johnston (2001) show, for example, that conversations with mobile phones, but not the mere holding of a mobile phone or listening to audiobooks, impair the recognition of traffic lights. Additional eye-tracking data indicate a lower attention performance regarding visual information as soon as a phone call is made simultaneously with a mobile phone (Strayer & Drews, 2007). In addition, studies show a connection between deficits in the divided attention of older adults and poorer performance in driving (De Raedt & Ponjaert-Kristoffersen, 2000), an increased involvement in accidents (Owsley et al., 1998), and a reduced recognition of traffic signs (Chaparro, Wood, & Carberry, 2005).

In addition to processes of selective and divided attention, the ability to switch tasks is often examined in the context of driving performance. Previous results, for example, demonstrate a correlation between performance in an attention-switching task and accident risk (Elander, West, & French, 1993; Mihal & Barrett, 1976). Furthermore, it is argued that slower reaction times when using a mobile phone with headphones compared to using it in hands-free mode (via speakers) are due to a required shift in spatial attention (Fagioli & Ferlazzo, 2006; Ferlazzo, Fagioli, Di Nocera, & Sdoia, 2008). Hunt, Morris, Edwards, and Wilson (1993) found that performance within a simple paper-pencil test for attention switching significantly correlates with the driving performance of healthy older people as well as people with mild senile dementia.

Another attention domain that seems to play an important role in driving is vigilance, which describes the sustained attention performance or alertness of the respective person. Results from driving studies suggest an impairment of driving performance with increasing hypovigilance (Larue, Rakotonirainy, & Pettitt, 2011). Furthermore, participants show an increased accident risk on monotonous roads, due to the loss of attention (Larue et al., 2011; Thiffault & Bergeron, 2003). Schmidt et al. (2007) show a linear deterioration of reaction times in this context.

To test attention performance in the context of driving, Ball, Owsley, Sloane, Roenker, and Bruni (1993) developed the "Useful Field of View Test" (UFOV), which includes three subtests of processing speed, divided attention, and selective attention (Ball et al., 1993; Ball & Rebok, 1994; Owsley, Ball, & Keeton, 1995; Owsley et al., 1998; Owsley, Ball, Sloane, Roenker, & Bruni, 1991). Ball et al. (1993) report a high sensitivity (89%) and specificity (81%) of the test in predicting the accident risk of older drivers. Clay et al. (2005) describe in their meta-analysis a large effect size (Cohen's $d = 0.945$) for the relationship between the UFOV test and negative driving outcomes in older people.

In summary, it can be said that previous driving studies often highlight the relevance of attention processes. However, these studies usually infer attention processes from the respective driving situations and the corresponding driving behavior. In addition, a domain-specific consideration within these studies was neglected. Although the UFOV test captures both selective and divided attention performance, it includes further visual and cognitive processing processes. The present study therefore aims to test the relationship between domain-specific attention performances and driving performance in a simulator. In order to investigate the respective attention areas as specifically as possible, tests were conducted on visual selective attention, auditory selective attention, visual divided attention, switching between different attention demands, switching between attributes, switching between rules, and vigilance. Based on earlier behavioral and neurophysiological studies showing a relationship between attention and inhibition processes as well as working memory (Awh & Jonides, 2001; Awh, Vogel, & Oh, 2006; Booth et al., 2003; Chun, 2011; Conway, Cowan, & Bunting, 2001; Lijffijt et al., 2009), three-way interaction effects with the respective attention areas and the two factors inhibition and working memory were tested (see Fig. 2). In addition, the aspect of age was taken into account by focusing on a wide age range.

The study was conducted on 123 participants ($M = 57.39$ years, $SD = 15.61$, 23-89 years, 31 women) of the total sample. The visual selective attention performance, the visual divided attention performance, and the switch between the two were tested with the "Switching Attentional Demands Task" (SwAD) (Liebherr, Antons, & Brand, 2019). The auditory selective attention performance was tested

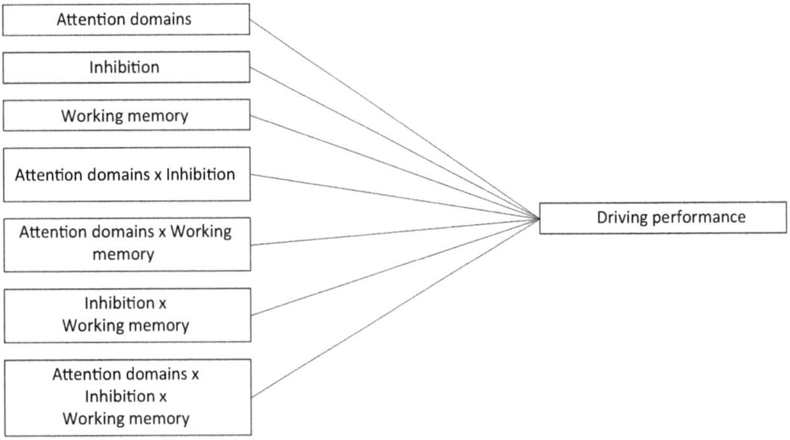

Fig. 2 Three-way interaction effects with the respective attention areas and the two factors inhibition and working memory

with a modified version of an Oddball task (Fichtenholtz et al., 2004). The Trail Making Test was used to test the switch between attributes (Reitan, 1971). In addition, the Modified Card Sorting Test was used to measure the switch between different rules (Nelson, 1976), the d2 Test to measure vigilance (Brickenkamp, 1962), the "Visual digit span backwards task" to quantify working memory performance (Ramsay & Reynolds, 1995) and an auditory "go/no-go-task" to test inhibition performance (Wegmann, Brand, Snagowski, & Schiebener, 2017), were used. The "Index of performance" described by Joshi et al. (2017) was used to measure driving performance. This was calculated after a standardized warm-up time of five minutes (IOP_{5-15}) and after an objective assessment of the adaptation point ($IOP_{adapt.}$). The objective assessment of the adaptation point was made by visually evaluating the "Index of performance" over time.

The mean values of the $IOP_{(5-15)}$ did not correlate significantly with specific attention areas as well as inhibition and working memory performance. Only 40 people showed an adaptation to the simulator. Additional analyses for these people revealed no significant relationships between $IOP_{(adapt)}$and the neuropsychological measurement methods used. Descriptive results and test statistics for the correlations are shown in Table 4. None of the three-way interaction models explained a significant portion of the variance (all $F(7, 122) \leq 1.47$, all $ps \geq 0.183$). Post-hoc, the correlations of age and the two driving performances were analyzed. Here, no significant correlations for the $IOP_{(5-15)}$ ($r = -0.076$, $p = 0.406$) and the $IOP_{(adapt)}$ ($r = -0.142$, $p = 0.384$), were found (Table 4).

4 Summary

Attention performance seems to be at least slightly related to adaptation to the simulator, with higher attention performance possibly favoring a (faster) adaptation, although this picture does not systematically show for all attention performances. The interplay of age and a) intelligence, b) cognitive flexibility, and c) attention did not lead to an accumulating effect on the timing of adaptation or the occurrence of simulator sickness. In line with previous studies, gender-specific differences in simulator sickness were identified. The results show a lower probability of simulator sickness occurring in individuals who have adapted to the simulator. The driving performance of the participants did not significantly correlate with the stress level before starting to drive or previous experiences with simulator sickness. However, age did significantly correlate with driving performance. Furthermore, individuals who had previously adapted to a simulator and those who had not showed no differences in cortisol levels prior to another simulator

Table 4 Correlations between driving performance and common measures for attention demands as well as inhibition and working memory

		M	*SD*	(1)	(2)	(3)	(4)	(5)	(6)	(7)	(8)	(9)	(10)	(11)
(1)	$IOP_{(5\text{-}15)}$	0.31	0.84	–										
(2)	$IOP_{(adapt.)}$	1.62	0.52	.246	–									
(3)	SwAD selective	457.07	59.31	−.109	.152	–								
(4)	SwAD divided	664.34	93.90	.024	.218	.613**	–							
(5)	SwAD Switch selective	473.65	58.98	−.146	.157	.736**	.556**	–						
(6)	SwAD Switch divided	648.58	87.18	−.057	.084	.685**	.747**	.681**	–					
(7)	Oddball	448.77	68.39	−.038	.204	.522**	.471**	.622**	.523**	–				
(8)	D2	149.94	46.40	.068	−.244	.130	.331**	.149	.309**	.174	–			
(9)	TMT (B)	75.08	28.64	.055	.053	.213*	.307**	.210*	.275**	.247**	.483**	–		
(10)	Digit span	5.20	1.52	−.095	−.204	−.015	−.156	−.012	−.120	−.114	−.317**	−.415**	–	
(11)	Go/Nogo-task	10.07	4.97	.021	−.006	.226*	.371**	.273**	.335**	.278**	.428**	.432**	−.374**	–
(12)	MCST	14.07	9.84	.157	.196	.133	.243**	.141	.175	.046	.345**	.439**	−.239**	.358**

All correlations with $IOP_{(adapt.)}$ refer to $N = 40$, $*p \leq .050$. $**p \leq .010$

drive. The results show no significant correlations between specific areas of attention and driving performance in the simulator. Three-way interactions of certain areas of attention with inhibition and working memory do not further clarify driving behavior in the simulator. The present results suggest using context-specific tasks for predicting driving performance rather than basic neuropsychological test procedures. Together with previous results, the relevance should be emphasized to distinguish between specific areas of attention such as selective attention or divided attention. With regard to the design and implementation of new vehicle technologies, we also recommend considering basic insights into specific effects such as modality and spatiality effects. However, more applied research is needed to better understand the effects of new vehicle technologies on attention processes and resulting driving performance and accident risk. In particular, with regard to the design of traffic safety programs, the relationships between specific areas of attention, driving performance, and age-associated changes should be considered from a more applied perspective in future research.

In addition, the result of a missing effect of previous experience with symptoms of simulator sickness on further driving performance in the simulator is of particular relevance for future simulator studies. However, this aspect should be further investigated with regard to different times between different measurement points and participants who had to abort the first simulator drive due to simulator sickness. Future studies on symptoms of simulator sickness should combine different approaches to measuring symptoms and further develop existing questionnaires taking into account the course of symptoms over time.

References

Al'Absi, M., Bongard, S., Buchanan, T., Pincomb, G. A., Licinio, J., & Lovallo, W. R. (1997). Cardiovascular and neuroendocrine adjustment to public speaking and mental arithmetic stressors. *Psychophysiology, 34*(3), 266–275.

Aldwin, C. M. (1991). Does age affect the stress and coping process? implications of age differences in perceived control. *Journal of gerontology, 46*(4), P174–P180.

Andrews, E. C., & Westerman, S. J. (2012). Age differences in simulated driving performance: Compensatory processes. *Accident Analysis & Prevention, 45*, 660–668.

Arthur Jr, W., & Doverspike, D. (1992). Locus of control and auditory selective attention as predictors of driving accident involvement: a comparative longitudinal investigation. *Journal of safety research, 23*(2), 73–80.

Awh, E., & Jonides, J. (2001). Overlapping mechanisms of attention and spatial working memory. *Trends in cognitive sciences, 5*(3), 119–126. https://doi.org/10.1016/s1364-6613(00)01593-x.

Awh, E., Vogel, E. K., & Oh, S. H. (2006). Interactions between attention and working memory. *Neuroscience, 139*(1), 201–208. https://doi.org/10.1016/j.neuroscience.2005.08.023.

Baldock, M. R. J., Mathias, J., McLean, J., & Berndt, A. (2007). Visual attention as a predictor of on-road driving performance of older drivers. *Australian Journal of Psychology, 59*(3), 159–168.

Balk, S. A., Bertola, M. A., & Inman, V. W. (2013). *Simulator sickness questionnaire: Twenty years later.* Paper presented at the Proceedings of the Seventh International Driving Symposium on Human Factors in Driver Assessment, Training, and Vehicle Design, Iowa City, IA, University of Iowa.

Ball, K., Owsley, C., Sloane, M. E., Roenker, D. L., & Bruni, J. R. (1993). Visual attention problems as a predictor of vehicle crashes in older drivers. *Investigative ophthalmology & visual science, 34*(11), 3110–3123.

Ball, K., & Rebok, G. (1994). Evaluating the driving ability of older adults. *Journal of Applied Gerontology, 13*(1), 20–38. https://doi.org/10.3758/pbr.17.1.15.

Bayarri, S., Fernandez, M., & Perez, M. (1996). Virtual reality for driving simulation. *Communications of the ACM, 39*(5), 72–76.

Booth, J. R., Burman, D. D., Meyer, J. R., Lei, Z., Trommer, B. L., Davenport, N. D., . . . Mesulam, M. M. (2003). Neural development of selective attention and response inhibition. *Neuroimage, 20*(2), 737–751. https://doi.org/10.1016/s1053-8119(03)00404-x.

Brickenkamp, R. (1962). *Aufmerksamkeits-Belastungs-Test (Test d2).* Göttingen: Hogrefe.

Brooks, J. O., Goodenough, R. R., Crisler, M. C., Klein, N. D., Alley, R. L., Koon, B. L., . . . Wills, R. F. (2010). Simulator sickness during driving simulation studies. *Accident Analysis & Prevention, 42*(3), 788–796. doi:doi.org/10.1016/j.aap.2009.04.013

Cabib, S., & Puglisi-Allegra, S. (2012). The mesoaccumbens dopamine in coping with stress. *Neuroscience & Biobehavioral Reviews, 36*(1), 79–89.

Cantin, V., Lavallière, M., Simoneau, M., & Teasdale, N. (2009). Mental workload when driving in a simulator: effects of age and driving complexity. *Accident Analysis & Prevention, 41*(4), 763–771.

Chaparro, A., Wood, J. M., & Carberry, T. (2005). Effects of age and auditory and visual dual tasks on closed-road driving performance. *Optometry and vision science, 82*(8), 747–754. https://doi.org/10.1097/01.opx.0000174724.74957.45.

Choukèr, A., Kaufmann, I., Kreth, S., Hauer, D., Feuerecker, M., Thieme, D., . . . Schelling, G. (2010). Motion sickness, stress and the endocannabinoid system. *PloS one, 5*(5), e10752. https://doi.org/10.1371/journal.pone.0010752.

Chun, M. M. (2011). Visual working memory as visual attention sustained internally over time. *Neuropsychologia, 49*(6), 1407–1409. https://doi.org/10.1016/j.neuropsychologia.2011.01.029.

Claremont, C. A. (1931). The psychology of seasickness. *Psyche, 11*, 86–90.

Classen, S., Bewernitz, M., & Shechtman, O. (2011). Driving simulator sickness: an evidence-based review of the literature. *American Journal of Occupational Therapy, 65*(2), 179–188.

Clay, O. J., Wadley, V. G., Edwards, J. D., Roth, D. L., Roenker, D. L., & Ball, K. K. (2005). Cumulative meta-analysis of the relationship between useful field of view and driving performance in older adults: Current and future implications. *Optometry and vision science, 82*(8), 724–731. https://doi.org/10.1097/01.opx.0000175009.08626.65.

Cobb, S. V., Nichols, S., Ramsey, A., & Wilson, J. R. (1999). Virtual reality-induced symptoms and effects (VRISE). *Presence: Teleoperators & Virtual Environments, 8*(2), 169–186. https://doi.org/10.1162/105474699566152.

Cohen, J. D., Braver, T. S., & Brown, J. W. (2002). Computational perspectives on dopamine function in prefrontal cortex. *Current opinion in neurobiology, 12*(2), 223–229.

Cohen, J. D., Cohen, P., West, S. G., & Aiken, L. S. (2003). *Applied multiple regression/correlation analysis for the behavioral sciences*. Mahwah, NJ: Lawrence Erlbaum.

Conway, A. R., Cowan, N., & Bunting, M. F. (2001). The cocktail party phenomenon revisited: The importance of working memory capacity. *Psychonomic Bulletin & Review, 8*(2), 331–335. https://doi.org/10.3758/bf03196169.

De Raedt, R., & Ponjaert-Kristoffersen, I. (2000). The relationship between cognitive/neuropsychological factors and car driving performance in older adults. *Journal of the American Geriatrics Society, 48*(12), 1664–1668. https://doi.org/10.1111/j.1532-5415.2000.tb03880.x.

Desmond, P. A., & Matthews, G. (1997). Implications of task-induced fatigue effects for in-vehicle countermeasures to driver fatigue. *Accident Analysis & Prevention, 29*(4), 515–523.

Dickerson, S. S., & Kemeny, M. E. (2004). Acute stressors and cortisol responses: a theoretical integration and synthesis of laboratory research. *Psychological Bulletin, 130*(3), 355.

Domeyer, J. E., Cassavaugh, N. D., & Backs, R. W. (2013). The use of adaptation to reduce simulator sickness in driving assessment and research. *Accident Analysis & Prevention, 53*, 127–132. https://doi.org/10.1016/j.aap.2012.12.039.

Dorn, L., & Matthews, G. (1995). Prediction of mood and risk appraisals from trait measures: Two studies of simulated driving. *European Journal of Personality, 9*(1), 25–42.

Ebenholtz, S. M. (1992). Motion sickness and oculomotor systems in virtual environments. *Presence: Teleoperators & Virtual Environments, 1*(3), 302–305. https://doi.org/10.1162/pres.1992.1.3.302.

Elander, J., West, R., & French, D. (1993). Behavioral correlates of individual differences in road-traffic crash risk: An examination of methods and findings. *Psychological Bulletin, 113*(2), 279. https://doi.org/10.1037/0033-2909.113.2.279.

Engström, J., Johansson, E., & Östlund, J. (2005). Effects of visual and cognitive load in real and simulated motorway driving. *Transportation Research Part F: Traffic Psychology and Behaviour, 8*(2), 97–120.

Eversmann, T., Gottsmann, M., Uhlich, E., Ulbrecht, G., Werder, K., & Scriba, P. C. (1978). Increased secretion of growth hormone, prolactin, antidiuretic hormone, and cortisol induced by the stress of motion sickness. *Aviation, space & environmental medicine*(1), 53–57.

Fagioli, S., & Ferlazzo, F. (2006). Shifting attention across spaces while driving: are hands-free mobile phones really safer? *Cognitive Processing, 7*(1), 147–147. https://doi.org/10.1007/s10339-006-0110-6.

Ferlazzo, F., Fagioli, S., Di Nocera, F., & Sdoia, S. (2008). Shifting attention across near and far spaces: Implications for the use of hands-free cell phones while driving. *Accident Analysis & Prevention, 40*(6), 1859–1864. https://doi.org/10.1016/j.aap.2008.07.003.

Fichtenholtz, H. M., Dean, H. L., Dillon, D. G., Yamasaki, H., McCarthy, G., & LaBar, K. S. (2004). Emotion–attention network interactions during a visual oddball task. *Cognitive Brain Research, 20*(1), 67–80. https://doi.org/10.1016/s0926-6410(04)00037-0.

Findlater, L., Froehlich, J. E., Fattal, K., Wobbrock, J. O., & Dastyar, T. (2013). *Age-related differences in performance with touchscreens compared to traditional mouse input.* Paper presented at the Proceedings of the SIGCHI Conference on Human Factors in Computing Systems.

Freund, B., & Green, T. R. (2006). Simulator sickness amongst older drivers with and without dementia. *Advances in Transportation Studies*, 71–74.

Garcia, A., Baldwin, C., & Dworsky, M. (2010). *Gender differences in simulator sickness in fixed-versus rotating-base driving simulator.* Paper presented at the Proceedings of the Human Factors and Ergonomics Society Annual Meeting.

Godley, S. T., Triggs, T. J., & Fildes, B. N. (2002). Driving simulator validation for speed research. *Accident Analysis & Prevention, 34*(5), 589–600.

Gulian, E., Matthews, G., Glendon, A. I., Davies, D. R., & Debney, L. M. (1989). Dimensions of driver stress. *Ergonomics, 32*(6), 585–602.

Horberry, T., Anderson, J., & Regan, M. A. (2006). The possible safety benefits of enhanced road markings: a driving simulator evaluation. *Transportation Research Part F: Traffic Psychology and Behaviour, 9*(1), 77–87. https://doi.org/10.1016/j.trf.2005.09.002.

Horberry, T., Anderson, J., Regan, M. A., Triggs, T. J., & Brown, J. W. (2006). Driver distraction: The effects of concurrent in-vehicle tasks, road environment complexity and age on driving performance. *Accident Analysis & Prevention, 38*(1), 185–191. https://doi.org/10.1016/j.aap.2005.09.007.

Horn, W. (1983). *L-P-S-Leistungsprüfungssystem.* Göttingen: Hogrefe.

Horrey, W. J., & Wickens, C. D. (2007). In-vehicle glance duration: distributions, tails, and model of crash risk. *Transportation research record, 2018*(1), 22–28. https://doi.org/10.3141/2018-04.

Hunt, L., Morris, J. C., Edwards, D., & Wilson, B. S. (1993). Driving performance in persons with mild senile dementia of the Alzheimer type. *Journal of the American Geriatrics Society, 41*(7), 747–753. https://doi.org/10.1111/j.1532-5415.1993.tb07465.x.

Jäncke, L., Brunner, B., & Esslen, M. (2008). Brain activation during fast driving in a driving simulator: the role of the lateral prefrontal cortex. *Neuroreport, 19*(11), 1127–1130.

Johnson, M. J., Chahal, T., Stinchcombe, A., Mullen, N., Weaver, B., & Bedard, M. (2011). Physiological responses to simulated and on-road driving. *International journal of Psychophysiology, 81*(3), 203–208.

Joshi, S. S., Maas, N., & Schramm, D. (2017). *A vehicle dynamics based algorithm for driver evaluation.* Paper presented at the 11th International Conference on Intelligent Systems and Control (ISCO).

Juster, R. P., Perna, A., Marin, M. F., Sindi, S., & Lupien, S. J. (2012). Timing is everything: Anticipatory stress dynamics among cortisol and blood pressure reactivity and recovery in healthy adults. *Stress, 15*(6), 569–577.

Kang, N. E., & Yoon, W. C. (2008). Age-and experience-related user behavior differences in the use of complicated electronic devices. *International Journal of Human-Computer Studies, 66*(6), 425–437.

Kawano, N., Iwamoto, K., Ebe, K., Aleksic, B., Noda, A., Umegaki, H., . . . Ozaki, N. (2012). Slower adaptation to driving simulator and simulator sickness in older adults aging clinical and experimental research. *Aging clinical and experimental research, 24*(3), 285–289.

Kemeny, A., & Panerai, F. (2003). Evaluating perception in driving simulation experiments. *Trends in cognitive sciences, 7*(1), 31–37.

Klauer, S. G., Dingus, T. A., Neale, V. L., Sudweeks, J. D., & Ramsey, D. J. (2006). The impact of driver inattention on near-crash/crash risk: An analysis using the 100-car naturalistic driving study data. *Springfield, Virginia: National Technical Information Service*. https://doi.org/10.1037/e729262011-001.

Klee, H., Bauer, C., Radwan, E., & Al-Deek, H. (1999). Preliminary validation of driving simulator based on forward speed. *Transportation research record, 1689*(1), 33–39.

Larue, G. S., Rakotonirainy, A., & Pettitt, A. N. (2011). Driving performance impairments due to hypovigilance on monotonous roads. *Accident Analysis & Prevention, 43*(6), 2037–2046. https://doi.org/10.1016/j.aap.2011.05.023.

Lee, H. C., Cameron, D., & Lee, A. H. (2003). Assessing the driving performance of older adult drivers: on-road versus simulated driving. *Accident Analysis & Prevention, 35*(5), 797–803.

Lerner, N., & Boyd, S. (2005). *On-road study of willingness to engage in distracting tasks*. Retrieved from

Lewis-Evans, B., & Charlton, S. G. (2006). Explicit and implicit processes in behavioural adaptation to road width. *Accident Analysis & Prevention, 38*(3), 610–617. https://doi.org/10.1016/j.aap.2005.12.005.

Lezak, M. (1995). *Neuropsychological Assessment (3rd edition)*. Oxford: Oxford University Press.

Liebherr, M., Antons, S., & Brand, M. (2019). The SwAD-Task–An Innovative Paradigm for Measuring Costs of Switching Between Different Attentional Demands. *Frontiers in Psychology, 10*, 2178.

Lijffijt, M., Lane, S. D., Meier, S. L., Boutros, N. N., Burroughs, S., Steinberg, J. L., . . . Swann, A. C. (2009). P50, N100, and P200 sensory gating: relationships with behavioral inhibition, attention, and working memory. *Psychophysiology, 46*(5), 1059–1068. https://doi.org/10.1111/j.1469-8986.2009.00845.x.

Lundqvist, A., Gerdle, B., & Rönnberg, J. (2000). Neuropsychological aspects of driving after a stroke—in the simulator and on the road. *Applied Cognitive Psychology: The Official Journal of the Society for Applied Research in Memory and Cognition, 14*(2), 135–150. https://doi.org/10.1002/(sici)1099-0720(200003/04)14:2<135::aid-acp628>3.0.co;2-s.

Mackrous, I., Lavallière, M., & Teasdale, N. (2014). Adaptation to simulator sickness in older drivers following multiple sessions in a driving simulator. *Gerontechnology, 12*(2), 101–111. https://doi.org/10.4017/gt.2013.12.2.004.00.

Matthews, G., Joyner, L. A., & Newman, R. (1999). *Age and gender differences in stress responses during simulated driving*. Paper presented at the Proceedings of the Human Factors and Ergonomics Society Annual Meeting.

McCartt, A. T., Hellinga, L. A., & Bratiman, K. A. (2006). Cell phones and driving: review of research. *Traffic Injury Prevention, 7*(2), 89–106.

McEwen, B. S. (1998). Protective and damaging effects of stress mediators. *New England journal of medicine, 338*(3), 171–179.

McEwen, B. S. (2010). Stress, sex and neural adaptation to a changing environment: mechanisms of neuronal remodeling. *Annals of the New York Academy of Sciences, 1204*(Suppl), E38.

McGehee, D. V., Lee, J. D., Rizzo, M., Dawson, J., & Bateman, K. (2004). Quantitative analysis of steering adaptation on a high performance fixed-base driving simulator. *Transportation Research Part F: Traffic Psychology and Behaviour, 7*(3), 181–196.

Mihal, W. L., & Barrett, G. V. (1976). Individual differences in perceptual information processing and their relation to automobile accident involvement. *Journal of applied psychology, 61*(2), 229. https://doi.org/10.1037//0021-9010.61.2.229.

Money, K. E., & Myles, W. S. (1974). Heavy water nystagmus and effects of alcohol. *Nature, 247*(5440), 404–405. https://doi.org/10.1038/247404a0.

Mullen, N. W., Weaver, B., Riendeau, J. A., Morrison, L. E., & Bédard, M. (2010). Driving performance and susceptibility to simulator sickness: Are they related? *American Journal of Occupational Therapy, 64*(2), 288–295. https://doi.org/10.5014/ajot.64.2.28810.5014/ajot.64.2.288.

Nelson, H. E. (1976). A modified card sorting test sensitive to frontal lobe defects. *Cortex, 12*(4), 313–324. https://doi.org/10.1016/s0010-9452(76)80035-4.

Otto, B., Riepl, R. L., Klosterhalfen, S., & Enck, P. (2006). Endocrine correlates of acute nausea and vomiting. *Autonomic Neuroscience, 129*(1-2), 17–21.

Owsley, C., Ball, K., & Keeton, D. M. (1995). Relationship between visual sensitivity and target localization in older adults. *Vision research, 35*(4), 579–587. https://doi.org/10.1016/0042-6989(94)00166-j.

Owsley, C., Ball, K., McGwin Jr, G., Sloane, M. E., Roenker, D. L., White, M. F., & Overley, E. T. (1998). Visual processing impairment and risk of motor vehicle crash among older adults. *Jama, 279*(14), 1083–1088. https://doi.org/10.1001/jama.279.14.1083.

Owsley, C., Ball, K., Sloane, M. E., Roenker, D. L., & Bruni, J. R. (1991). Visual/cognitive correlates of vehicle accidents in older drivers. *Psychology and aging, 6*(3), 403. https://doi.org/10.1037//0882-7974.6.3.403.

Owsley, C., & McGwin Jr, G. (2010). Vision and driving. *Vision research, 50*(23), 2348–2361.

Ramos, B. P., & Arnsten, A. F. (2007). Adrenergic pharmacology and cognition: focus on the prefrontal cortex. *Pharmacology & therapeutics, 113*(3), 523–536.

Ramsay, M. C., & Reynolds, C. R. (1995). Separate digits tests: A brief history, a literature review, and a reexamination of the factor structure of the Test of Memory and Learning (TOMAL). *Neuropsychology Review, 5*(3), 151–171. https://doi.org/10.1007/bf02214760.

Reason, J. T. (1978). Motion sickness adaptation: a neural mismatch model. *Journal of the Royal Society of Medicine, 71*(11), 819–829. https://doi.org/10.1177/014107687807101109.

Reimer, B., Mehler, B. L., Pohlmeyer, A. E., Coughlin, J. F., & Dusek, J. A. (2006). The use of heart rate in a driving simulator as an indicator of age-related differences in driver workload. *Advances in Transportation Studies*, 9–29.

Reitan, R. M. (1971). Trail making test results for normal and brain-damaged children. *Perceptual and motor skills, 33*(2), 575–581. https://doi.org/10.2466/pms.1971.33.2.575.

Richardson, E. D., & Marottoli, R. A. (2003). Visual attention and driving behaviors among community-living older persons. *The Journals of Gerontology Series A: Biological Sciences and Medical Sciences, 58*(9), M832–M836. https://doi.org/10.1093/gerona/58.9.m832.

Roupa, Z., Nikas, M., Gerasimou, E., Zafeiri, V., Giasyrani, L., Kazitori, E., & Sotiropoulou, P. (2010). The use of technology by the elderly. *Health Science Journal, 4*(2), 118.

Sahami, S., Jenkins, J. M., & Sayed, T. (2009). Methodology to analyze adaptation in driving simulators. *Transportation research record, 2138*(1), 94–101.

Stanney, K. M., Kingdon, K. S., & Kennedy, R. S. (2002). Dropouts and Aftereffects: Examining General Accessibility to Virtual Environment Technology. *Proceedings of the Human Factors and Ergonomics Society Annual Meeting, 46*(26), 2114–2118. https://doi.org/10.1177/154193120204602603.

Stanney, K. M., Mourant, R. R., & Kennedy, R. S. (1998). Human Factors Issues in Virtual Environments: A Review of the Literature. *Presence: Teleoperators and Virtual Environments, 7*(4), 327–351. https://doi.org/10.1162/105474698565767.

Sterling, P. (2012). Allostasis: a model of predictive regulation. *Physiology & behavior, 106*(1), 5–15.

Sterling, P., & Eyer, J. (1988). Allostasis: a new paradigm to explain arousal pathology. In S. Fisher & J. Reason (Eds.), *Handbook of life stress, cognition and health* (pp. 629–649): J. Wiley and Sons: New York.

Strayer, D. L., & Drews, F. A. (2007). Cell-phone-induced driver distraction. *Current Directions in Psychological Science., 16*(3), 128–131. https://doi.org/10.1111/j.1467-8721.2007.00489.x.

Strayer, D. L., & Johnston, W. A. (2001). Driven to distraction: dual-task studies of simulated driving and conversing on a cellular telephone. *Psychological Science, 12*(6), 462–466. https://doi.org/10.1111/1467-9280.00386.

Takayama, L., & Nass, C. (2008). Driver safety and information from afar: an experimental driving simulator study of wireless vs. in-car information services. *International Journal of Human-Computer Studies, 66*(3), 173–184. https://doi.org/10.1016/j.ijhcs.2006.06.005.

Thiffault, P., & Bergeron, J. (2003). Monotony of road environment and driver fatigue: a simulator study. *Accident Analysis & Prevention, 35*(3), 381–391. https://doi.org/10.1016/s0001-4575(02)00014-3.

Uchiyama, Y., Ebe, K., Kozato, A., Okada, T., & Sadato, N. (2003). The neural substrates of driving at a safe distance: a functional MRI study. *Neuroscience letters, 352*(3), 199–202.

Underwood, G., Crundall, D., & Chapman, P. (2011). Driving simulator validation with hazard perception. *Transportation Research Part F: Traffic Psychology and Behaviour, 14*(6), 435–446.

Vrijkotte, T. G., Van Doornen, L. J., & De Geus, E. (2000). Effects of work stress on ambulatory blood pressure, heart rate, and heart rate variability. *Hypertension, 35*(4), 880–886.

Wecker, N. S., Kramer, J. H., Hallam, B. J., & Delis, D. C. (2005). Mental flexibility: age effects on switching. *Neuropsychology, 19*(3), 345. https://doi.org/10.1037/0894-4105.19.3.345.

Wegmann, E., Brand, M., Snagowski, J., & Schiebener, J. (2017). Are you able not to react to what you hear? Inhibition behavior measured with an auditory Go/NoGo paradigm. *Journal of clinical and Experimental Neuropsychology, 39*(1), 58–71. https://doi.org/10.1080/13803395.2016.1201461.

Weinberg, G., & Harsham, B. (2009). Developing a low-cost driving simulator for the evaluation of in-vehicle technologies. *AutomotiveUI, 9*, 51–54.

Wikman, A. S., Nieminen, T., & Summala, H. (1998). Driving experience and time-sharing during in-car tasks on roads of different width. *Ergonomics, 41*(3), 358–372. https://doi.org/10.1080/001401398187080.

Wood, J. M. (2002). Age and visual impairment decrease driving performance as measured on a closed-road circuit. *Human factors, 44*(3), 482–494.

Yamaguchi, M., & Sakakima, J. (2007). Evaluation of driver stress in a motor-vehicle driving simulator using a biochemical marker. *Journal of international medical research, 35*(1), 91–100.

Market Potentials of Older Drivers - Willingness to Pay and Acceptance of Age-Appropriate Driver Assistance Systems

Timo Günthner, Lukas Zeymer, Heike Proff and Josip Jovic

Contents

Dr. Timo Günthner, Lukas Zeymer, M.Sc., Prof. Dr. Heike Proff, Dr. Josip Jovic, all University of Duisburg-Essen

T. Günthner · L. Zeymer · H. Proff · J. Jovic (✉)
Lehrstuhl für ABWL & Internationales Automobilmanagement, Universität Duisburg-Essen, Duisburg, Germany
e-mail: josip.jovic@uni-due.de

T. Günthner
e-mail: timo.guenthner@uni-due.de

L. Zeymer
e-mail: lukas.zeymer@uni-due.de

H. Proff
e-mail: heike.proff@uni-due.de

H. Proff et al. (eds.), *Age-appropriate Driver Assistance Systems*,
https://doi.org/10.1007/978-3-658-44838-7_9

In light of the stagnating automobile markets in the triad, automobile companies are looking for new market potentials, e.g., the growing group of older drivers (Proff 2019). As shown in the Chapter "Mobility in Old Age - An Introduction" of this book, their number is increasing. They have, at least in more developed countries, money for offers that increase safety and comfort and allow them to remain mobile for longer in old age despite physical and psychological limitations. From a management perspective, it is therefore important to know whether market potentials exist for age-appropriate driver assistance systems, i.e., above all, whether older drivers are willing to pay for more mobility and thus the preservation of their mobility in old age. Only then can profitable business models be developed.

For the five driver assistance systems selected in the preliminary examination (see Chapter "Older Drivers: Investigation and Sample in the ALFASY Project"), namely cross-traffic alert, parking distance control, blind spot detection, lane departure warning, and tire pressure monitoring system, the willingness to pay was examined depending on age and acceptance, which is considered a major influencing factor on the willingness to pay. In the first of the three main examinations, willingness to pay and acceptance of the offered assistance systems were initially recorded in the driving simulator. Subsequently, improved acoustic assistance systems were tested, in the second main examination also in the simulator and in the third main examination after a test drive. Since age and life events do not always coincide, the influence of critical life events in old age such as illness or death of the spouse or life partner was also asked.

This chapter first presents important research findings (Section 1), then seeks explanations for the willingness to pay and acceptance of driver assistance systems for older drivers as well as for the influence of critical life events (Section 2). In Section 3, the research concept is explained. Subsequently, the research results of the first main examinaitons on the willingness to pay and acceptance of the offered driver assistance systems are summarized (Section 4) and compared with the results of the second main examinaitonon improved acoustic driver assistance systems (Sect. 5). The results of the third main examination after a drive on a test track are only briefly mentioned, as the sample (see Chapter "Older Drivers: Investigation and Sample in the ALFASY Project") is relatively small (n = 46) and hardly comparable with the previous studies. Finally, conclusions from the ALFASY project for further research are discussed (Section 6).

1 Literature on Market Potentials, Acceptance, and Critical Life Events in Old Age

As evidenced in Chapter "Mobility in Old Age—An Introduction", mobility often stands for quality of life in old age (Burghard 2005; Engeln 2001). Through a car, older people can maintain social relationships and contacts, remain independent, and improve well-being (e.g., Nobis et al. 2019; Musselwhite et al. 2015; Edwards et al. 2009). Accordingly, perceived quality of life decreases when mobility is restricted (Musselwhite & Haddad 2018; Holley-Moore & Creighton 2015). Assistance systems are intended to help maintain or regain driving ability (Günthner et al. 2020; Boot & Scialfa 2016; Karthaus et al. 2016).

Although the importance of assistance systems and especially autonomous driving is known, there is little work on this in economic and social science research (Günthner et al. 2020; Musselwhite et al. 2015; Winner & Schopper 2015). However, there are some studies that demonstrate a higher willingness to pay among older people for products and services that improve quality of life (Souders et al. 2017; Schulz et al. 2013). Willingness to pay is understood as the purchase price that a customer is willing to pay at most for a product or service (Proff & Fojcik 2014; Miller et al. 2011) and is therefore considered an indicator of market potentials.

The few studies on willingness to pay for driver assistance systems come to different conclusions. According to Blythe and Curtis (2004), willingness to pay is low; according to a more recent study by BCG and MEMA (2015), it even falls below the current price for such systems. Trübswetter (2015), on the other hand, demonstrates a higher willingness to pay among those over 50, and thus a market potential. This result is confirmed by a study on the blind spot detection (Souders et al. 2017): Older people are willing to pay about twice as much for this assistant as young drivers, especially if they know it (Trübswetter & Bengler 2011 and Piao et al. 2005referring to the cruise control).

Other studies, in turn, show that acceptance is particularly high among older drivers (Son et al. 2015; Stevens 2012) and the evaluation is more positive than by younger drivers (Viborg 1999). They, however, view autonomous driving vehicles with skepticism (König & Neumayr 2017).

Elder (1994) was able to show that age does not linearly affect needs (see also Glenn 2003) and older people must be differentiated by age and by critical life events (see also Günthner et al. 2020). In the ALFASY project, therefore, willingness to pay and acceptance of age-appropriate driver assistance systems were examined in general and differentiated by age groups and critical life events, in order to estimate the market potential of assistance systems for older drivers.

2 Explanation and Assumptions about the Market Potentials of Driver Assistance Systems for Older Drivers

For the examniation of the market potentials of driver assistance systems for older drivers, i.e., the willingness to spend money to remedy perceived deficits and thus satisfy needs (Chapter "Older Drivers: Investigation and Sample in the ALFASY Project"), explanations for the willingness to pay for age-appropriate driver assistance systems are sought (Section 2.1). But also explanations of the acceptance of technical systems, each depending on age. Acceptance (Section 2.2) and being affected by critical life events (Section 2.3) are likely to influence willingness to pay.

2.1 Explanation of Market Potentials (Willingness to Pay) for Driver Assistance Systems in Older Drivers

The explanation of the market potentials of driver assistance systems for older drivers is based on household theory and explanations of intertemporal consumption (Proff 2019). According to the theory of the household, a household, like an individual, can optimize current and future consumption depending on income (Varian 2014, Chapter "Driver assistance systems in the context of age-appropriate HMI design"). Since the assumption that they are fully informed is unrealistic, baskets of goods and services are distinguished, e.g., everyday goods, consumer goods, and luxury goods, which change with age (see Fig. 1).

Studies show a hump-shaped progression of the consumption expenditure curve over the entire life cycle (e.g., Yang 2009; Fernández-Villaverde & Krueger 2007). The theory of intertemporal consumption (Blundell et al. 1994), which explains the relationship between age and demand across baskets of goods, confirms the results. It is assumed that a household determines its consumption expenditures in such a way that the marginal utility of income remains constant over the course of life. This means that, for example, young people first meet their basic needs and in middle age, when demand is saturated, satisfy additional and less rational needs. They can then more realistically assess expected future incomes, income risks, and propensity to save (Proff 2019 based on Miles 1997). According to the uneven distribution of income and consumption curves over the life cycle (see also Fig. 1), borrowing, loan repayment, and savings are also unevenly distributed over the life cycle.

From the intertemporal consumption behavior of households with age-related different needs, baskets of goods can be derived (e.g., Moschis 2012 and Fig. 1):

- Basket of goods I of young households (basic needs),
- Basket of goods II of an average middle-aged family with young children (basic and luxury needs, especially middle-aged people) and
- Basket of goods III of retiree households (replacement goods, especially for older people).

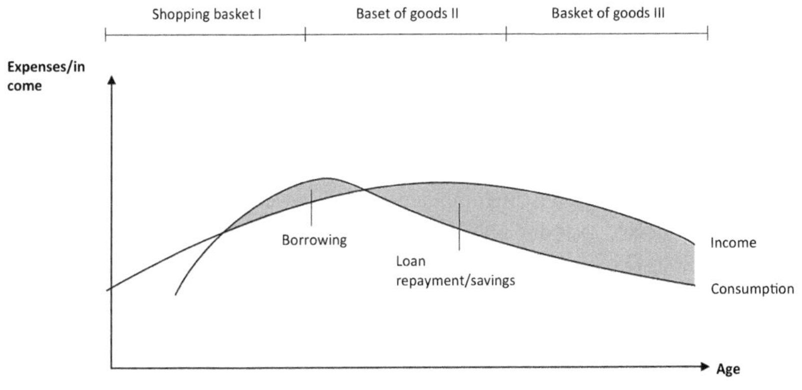

Fig. 1 Mobility needs (consumption) depending on the ages of consumers *Source* Proff 2019 based on Blundell et al. 1994

The population cohorts correspond - at least in Germany - roughly to the age groups under 25, 25 to 65, and over 65. However, this age distribution shows considerable country-specific differences, as both the mobility options and the degrees of freedom of mobility demand are different in individual countries (Proff 2019). Even within the group of older people (over 65 with basket of goods III), needs and consumer behavior vary.

Figure 1 can be extended because it can be assumed that older people not only buy goods to cover their replacement needs. It can be assumed (Section 1) that

- they have a higher willingness to pay for goods that improve quality of life (Schulz et al. 2013) and
- the willingness to pay for driver assistance systems increases with age, as they improve quality of life (Souders et al. 2017; Trübswetter 2015; Davidse 2006).

This justifies two hypotheses for the empirical investigation:

H_1 The willingness to pay for age-appropriate driver assistance systems correlates with age.

H_2 The willingness to pay for age-appropriate driver assistance systems increases with the average age of the age groups

2.2 Explanation of the Acceptance of Driver Assistance Systems among Older Drivers as a Factor Influencing Willingness to Pay

The willingness to pay for driver assistance systems increases when technologies for improving safety are socially desirable (Schleiffer 2024). Technology acceptance models attempt to explain the acceptance of new products and services as well as their use or non-use (e.g., Dudenhöffer 2015; Proff & Fojcik 2014).

Business research refers to computer science, psychology, and sociology in a variety of models to examine acceptance (e.g., Venkatesh & Bala 2008; Venkatesh & Davis 2000; Davis et al. 1989). They theoretically justify the intention to use products and services (Ajzen & Fishbein 1980). After an overview of common explanations of acceptance (1), the model for investigating the acceptance of new technologies in the ALFASY project is described (2).

(1) Explanations of Acceptance

The basis of many acceptance models are the Theory of Reasoned Action, the Theory of Planned Behavior, and a Technology Acceptance Model.

The Theory of Reasoned Action (TRA) by Ajzen and Fishbein (1980) explains actual behavior from a behavioral intention or usage intention, which results from the attitude towards behavior and a "subjective norm" as the perception of social pressure (Ajzen & Fishbein 1980 and Fig. 2). It was extended by the Theory of Planned Behavior (TPB), which, in addition to attitude and subjective norm, includes "perceived behavioral control" as a third influencing factor on the behavioral intention, which interacts with the other influencing factors (Ajzen 1991 and Fig. 2). It is assumed that a behavior is more likely the greater the subjective conviction is to have enough opportunities and resources, including skills, to realize the behavioral intention (see e.g., Dudenhöffer 2015).

Davis et al. (1989) developed a Technology Acceptance Model (TAM 1) based on the Theory of Planned Behavior, which sees "perceived usefulness" and "perceived ease of use" as the main influencing factors on usage intention and thus on usage behavior. With additions, this model forms the basis of many studies on the acceptance of new technologies or technological innovations. The Technology Acceptance Model 2 (TAM 2, Venkatesh & Davis 2000) is common, which considers subjective norm, image, work relevance, quality and visibility of the result as influencing factors on "perceived usefulness" and "experience" and "voluntariness of use" as moderator variables (Fig. 3).

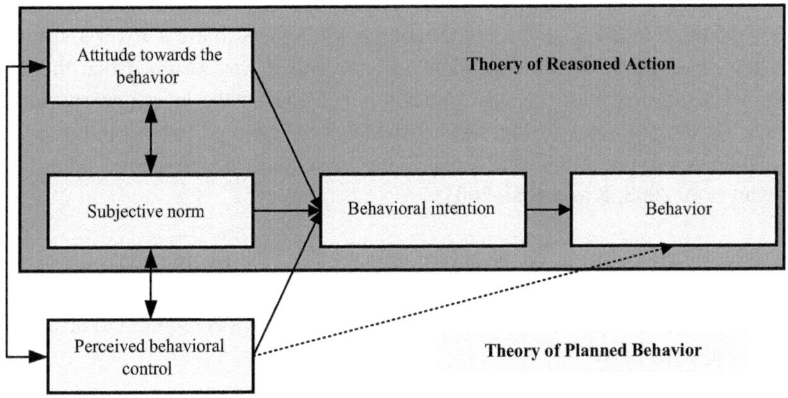

Fig. 2 Theory of Reasoned Action and Theory of Planned Behavior *Source* based on Ajzen & Fishbein 1980

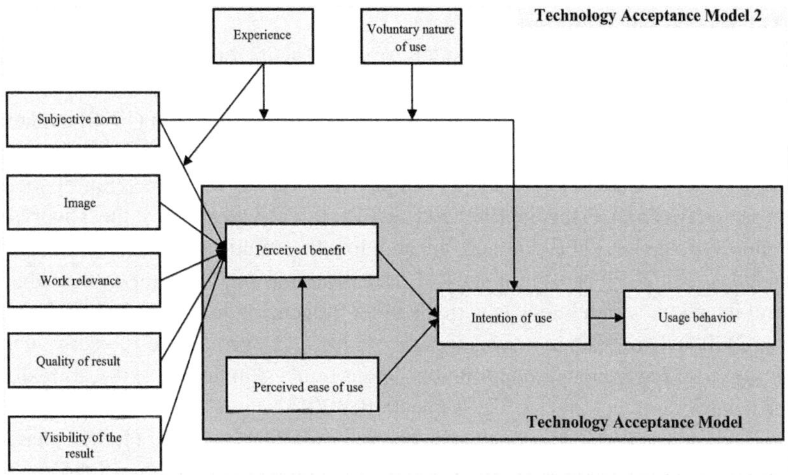

Fig. 3 Technology Acceptance Model 2 *Source* Venkatesh, Davis 2000

(2) Examination model of the acceptance of age-appropriate driver assistance systems in the ALFASY project
The Technology Acceptance Model 2 (TAM 2) forms the basis of the model for examining the acceptance of age-appropriate driver assistance systems in the ALFASY project (Fig. 4), because it allows a multitude of influencing factors on the intention to use and can therefore be easily adapted. As in the TAM 2 model, the "intention to use", i.e. the strength of the willingness to use a driver assistance system, was adopted as a dependent variable and it was assumed that this can predict actual behavior (cf. e.g. Dudenhöffer 2015). In the investigation model, three further variables of the TAM 2 model were adopted, whose influence on the intention to use and thus acceptance has been confirmed in many studies (e.g. Fagan et al. 2008; Kwon et al. 2007):

1. "Subjective norm" as the perceived pressure from the environment on a person to use or not to use a particular system.
2. "Perceived ease of use" as the effort a person believes is required to be able to use a system and
3. "Perceived benefit" of the system for an activity.

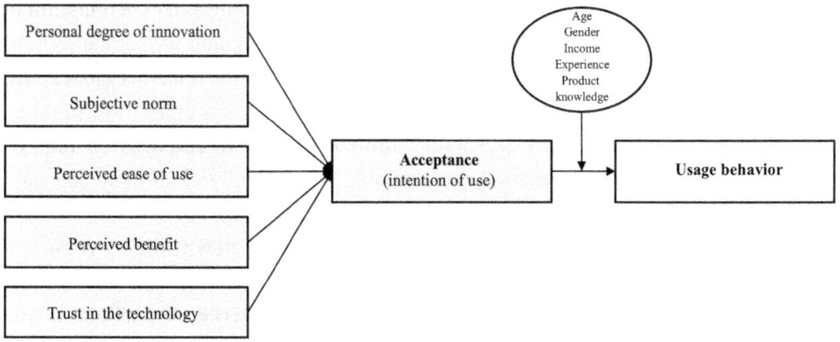

Fig. 4 Acceptance model of the examination in the ALFASY project

For the determination of the acceptance of age-appropriate driver assistance systems, two more variables were added, whose importance was shown in the preliminary examinaiton (Chapter "Older drivers: Investigation and sample in the ALFASY project"):

4. "Personal degree of innovation" as the enthusiasm to test a technology and innovation early on (cf. Agarwal & Prasad 1998; Midgley & Dowling 1978).
5. "Trust in technology" as the willingness to entrust oneself to a technology despite risks in expectation of a positive result, i.e. a benefit, when, for example, an assistance system is installed in a vehicle (Mayer et al. 1995 and on the influence on acceptance e.g. Kaur & Rampersad 2018).

Moderating variables of the ALFASY examination are, besides gender, income, experience and product knowledge, especially the age of the respondents (cf. Chapter "Older drivers: Investigation and sample in the ALFASY project"), which is increasingly taken into account in acceptance models in view of demographic change. According to a study by Souders and Charness (2016) on the attitude of older people (55 years and older) towards driver assistance systems, the willingness to use them increases with age. Braun et al. (2019) can show that acceptance is particularly high among those over 65, with assistance systems with a lower degree of autonomy being more accepted.

The examination model (Fig. 4) assumes that with increasing age, acceptance and willingness to pay for age-appropriate driver assistance systems increase.

H_3 The acceptance of age-appropriate driver assistance systems correlates with age.

H_4 The acceptance of age-appropriate driver assistance systems *increases* with the average age of the age groups.

H_5 Acceptance significantly positively influences the willingness to pay for age-appropriate driver assistance systems

2.3 Explanation of the Influence of Critical Life Events on the Willingness to Pay and Acceptance of Driver Assistance Systems by Older Drivers

Not only income, but also critical events in the course of life can influence purchasing behavior, so that age does not linearly determine needs and demand (cf. Elder 1994). Therefore, critical events in the lives of older people should be included in an examination of age-appropriate driver assistance systems (cf. Horlacher 2000; Filipp 1995).

Such investigations are based on psychological research. They are mostly based on the life course theory today (Elder 1994; Fuchs-Heinritz 2009). Empirically, the assumption of a psychosocial causation of mental and physical illnesses is tested and it is assumed that critical life events, which are unpredictable, can contribute to mental disorders (cf. Filipp 1995).

Critical events do not occur at a certain age, but increase with age. A distinction is made between age-standardized events such as studying, starting a career, starting a family, and ending professional activity (cf. e.g. Montada 2008) and non-age-standardized events such as unemployment, life-threatening diseases, death of a partner or relatives (cf. Fig. 5).

Life events such as the birth of a child, separation, death of a life partner or diseases influence the choice of transport (cf. Chatterjee & Scheiner 2015; Müggenburg & Lanzendorf 2015), the ownership of a vehicle (cf. Clark et al. 2016) and also the use of car sharing (cf. Uteng et al. 2019).

The ALFASY project examined the end of professional activity, the death of a partner, the birth of the first grandchild, a traffic accident (whether self-inflicted or caused by others), and a serious or chronic illness as relevant events for the willingness to pay and acceptance of driver assistance systems:

H_6 Acceptance and willingness to pay for age-appropriate driver assistance systems are influenced by the following critical life events: a) end of professional activity, b) death of a partner, c) birth of the first grandchild, d) self-inflicted traffic accident, e) traffic accident caused by others, f) serious illness and g) chronic illness.

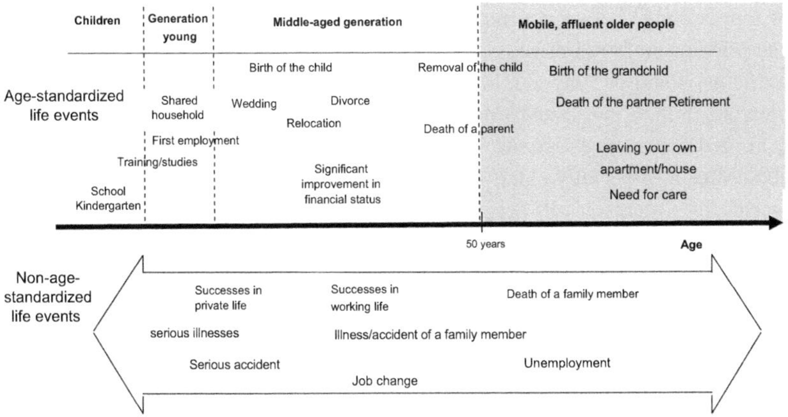

Fig. 5 Critical events in the course of life *Source* Proff 2019 based on Montada 2008; Elder 1994

3 Concept of Investigating Acceptance and Willingness to Pay

The hypotheses of increasing willingness to pay (H_1 and H_2) and acceptance (H_3 to H_5) for age-appropriate driver assistance systems with increasing age and with the rise of the average age of the age groups were empirically tested. The samples of the studies are presented in Chap. "Older Drivers: Investigation and Sample in the ALFASY Project". The operationalization of the variables (Sect. 3.1) and the research approach (Sect. 3.2) are presented below.

3.1 Operationalization of Variables

As mentioned, willingness to pay is defined by the price a customer is willing to pay at most for a product or service (cf. Proff 2019; Miller et al. 2011). The price query is made directly via individual partial utility values (e.g., Proff et al. 2018). In the ALFASY project, a method was chosen to determine the willingness to pay for driver assistance systems that is based on the approach of van Westendorp (1976) (cf. also Kloss & Kunter 2016).

It can be assumed that potential buyers have an idea of how much they are willing to pay for a driver assistance system, but cannot specify an exact price or a price range between a minimum and a maximum price. Therefore, no average price or preferred price for the five selected driver assistance systems (cross-traffic assistant, parking aid, blind spot assistant, lane keeping assistant, and tire pressure monitoring system) of the preliminary study can be determined (Proff et al. 2018; Simon & Fassnacht 2016). The subjects were therefore not asked about this, as especially older people will probably not know some or all of these assistance systems at all, because they are not installed in their vehicles and a possible surcharge is not known to them. Therefore, they have to go through a learning process with new knowledge (cf. e.g., Leifer et al. 2001). Since a derivation of "optimal prices" is also not possible (cf. Roll et al. 2018), the van Westendorp method was slightly modified: Instead of an open price statement with the characteristics "too expensive" or "expensive", "cheap" or "too cheap", the subjects were asked to evaluate predefined price ranges. These were defined in consultation with the project partner Ford (e.g., for a cross-traffic assistant less than 290 €, 290 to 580 €, 580 to 870 € and more than 870 €). The willingness to pay for a specific driver assistance system was then determined by aggregating the responses (Proff 2019) to a total of nine items (price specifications), three of which were for a visual, three for a visual-haptic, and three for a visual-acoustic variant of the five driver assistance systems.

The acceptance (intention to use as a dependent variable) of the five driver assistance systems was recorded via five independent variables (cf. Sect. 2.2 (2)): 1) subjective norm, 2) perceived ease of use, 3) perceived benefit, 4) personal degree of innovation, and 5) trust in technology. The questionnaire contains all six variables with a total of 14 items (Table 1) on a 7-point Likert scale from (1) "strongly disagree" to (7) "strongly agree".

The moderating variables gender, age, income, experience, product knowledge, as well as the critical life events a) end of professional activity, b) death of the partner, c) birth of the first grandchild, d) self-inflicted serious accident, e) serious accident caused by others, f) serious illness, and g) chronic illness were also recorded in the preliminary study on a 7-point Likert scale from 1 (very low) to 7 (very high).

3.2 Research Approach

First, the relationship between age and willingness to pay for age-appropriate driver assistance systems (H_1) and the significance of the differences between the three age groups of older people (H_2) were investigated, then the relationship between age and acceptance of driver assistance systems (H_3) and again

Table 1 Operationalization of variables

a) Operationalization of the independent variables

Variable	Operationalization		Source
Personal degree of innovation	PERIN_01	I am open to new products I find it interesting to try out new products	Lee (2019); Königstorfer (2008); Rogers (2003); Parasuraman (2000)
	PERIN_02	I find it interesting to try out new products	
	PERIN_03		
	PERIN_04	I am usually the one who informs others about new products	
Subjective norm	SN_01	People who are important to me would appreciate it if I used the driver assistance system	Taylor and Todd (1995); Venkatesh and Davis (2000); Venkatesh et al. (2003)
	SN_02	People who influence my behavior would appreciate it if I used the driver assistance system	
Perceived ease of use	WEDN_01	It would be easy for me to learn how to use the driver assistance system	Davis et al (1989); Venkatesh et al. (2003)
	WEDN_02	I don't see any major difficulties in operating the driver assistance system	
Perceived benefit	WN_01	The driver assistance system is a good idea	Ajzen(1991); Davis et al (1989); Taylor and Todd (1995); Venkatesh and Davis (2000); Venkatesh et al. (2003)
	WN_02	I would find the driver assistance system useful	
Trust in the technology	VER_01	The driver assistance system is technically reliable	Gefen et al (2003); Pavlou (2003); Saeed et al (2003); McKnight et al. (2011)
	VER_02	I can trust the driver assistance systems	

b) Operationalization of the dependent variables

Variable	**Operationalization**		**Source**
Willingness to pay	Driver assistance system	I consider the price of less than 290 Eur. for the driver assistance system as "too expensive", "expensive", "cheap" or "too cheap "too cheap".	Van Westendorp (1976)
	Driver assistance system	I consider the price of 290 to 580 Eur. for the driver assistance system as "too expensive", "expensive", "cheap", too "cheap"	
	Driver assistance system	I consider the price of 580 Eur. to 870 Eur. for a driver assistance system as "too expensive", "expensive", "cheap" and "too cheap".	
	Driver assistance system	I consider a price of more than 870 Eur. for a driver assistance system as "too expensive", "expensive", "cheap" or "too chap".	
Intended use	NI_01I	I can image to use the driver assistance system	Davis et al (1989); Venkatesh et al. (2003)
	NI_02Having a driver assistance system	I can imagine that in my vehicle	

the significance of the differences between the age groups of older people (H_4). Finally, the relationship between acceptance and willingness to pay was examined (H_5) and the influence of critical life events (H_6) was investigated. Figure 6 shows the research model.

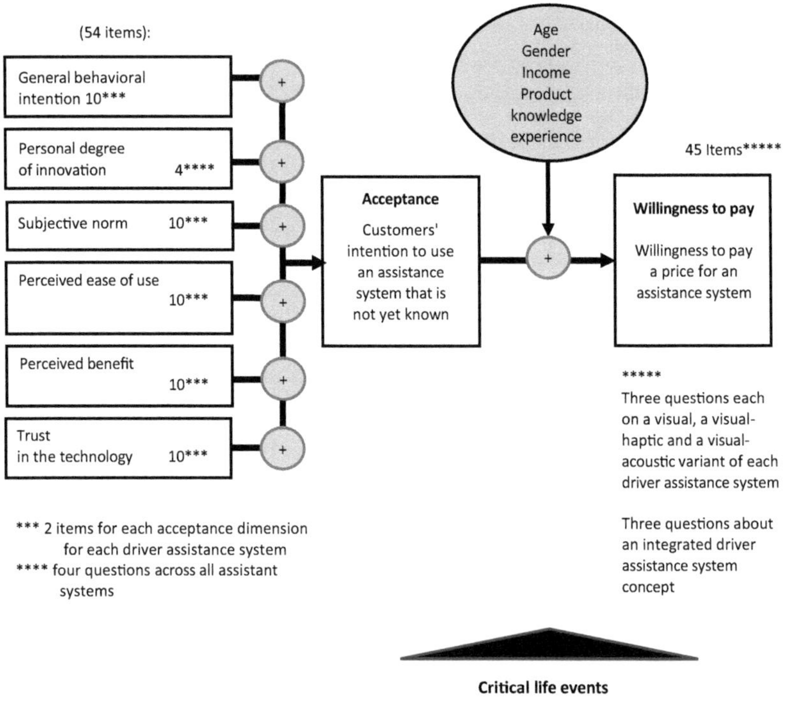

Fig. 6 Research model of the influencing factors on Willingness to pay

4 Results of the Examination of Market Potentials of Age-Appropriate Driver Assistance Systems for Older Drivers (First Main Examination)

The first main examination with 181 subjects aged 50 years and older took place in 2018 in a driving simulator at the University of Duisburg-Essen. Nearly 80% were male, the average age was 65.4 years (SD = 7.945). 27% were between 50 and 59 years old, 42% between 60 and 69 years old, and the rest over 70 years old, of which more than half were between 70 and 74 (see chap. "Older Drivers: Investigation and Sample in the ALFASY Project"). Most subjects had an income between 2,000 and 3,000 € per month. Experiences with age-appropriate driver assistants and product knowledge varied greatly. A comparison group of 46 younger drivers (under 50 years) was surveyed.

4.1 Results of the Examination of Willingness to Pay

First, the results of the correlation between age and willingness to pay for age-appropriate driver assistance systems are shown (1), then differences in willingness to pay between age groups (2).

(1) Correlation between Age and Willingness to Pay

Age and willingness to pay of older drivers correlate only to a limited extent. The correlation is significant for the blind spot assistant ($r(181) = .182$, $p < .01$) and tire pressure monitoring system ($r(181) = .240$, $p < .001$). The correlation is not significant for the cross-traffic alert, park distance control, and lane departure warning, i.e., the hypothesis H_1 can only be confirmed for two of the five driver assistance systems (Table 2). Of the control variables, only income has a significant influence on willingness to pay, experience and product knowledge do not. Willingness to pay increases with increasing income.

These results are consistent with the observation by Elder (1994) (Sections 1 and 2.3) that there is no clear linear relationship between age and need. They support the necessity of examining age groups of older drivers.

(2) Differences in Willingness to Pay Between Age Groups

Using a two-sample t-test for independent samples, differences in willingness to pay between the three age groups (50–59 years, 60–69 years, and 70 years and older) were to be identified at a significance level of $p \leq 0.05$ (highly significant) and $p \leq 0.10$ (slightly significant). All age groups were compared with each other (Table 3). The equality of variance was determined at the same significance

Table 2 Correlation Table Willingness to pay (n = 181)

	1. Age	2. Exper.	3. Prod. knowl.	4. Net in.	5. TWP CTA	6. WTP PDC	7. WTP BSD	S. WTP LDW	9. WTP TPMS
1. Age									
2. Experience	,029								
3. Product knowledge	,024	**,708****							
4. Net income	-,058	**,241**	**,363****						
5. WTP CTA	.057	.051	,005	,005					
6. WTP PDC	,126	,060	-,057	,087	**,652****				
7. WTP BSD	,182*	.121	,025	,116	**,645****	**,779**			
8. WTP LDW	,100	,022	,014	,123	**,680****	**,675****	**,634****		
9. WTP TPMS	,240**	,064	,020	,074	**,494****	**,629****	**,588****	**,603****	

*Correlation is significant at 0.01 level (two-sided)

****Correlation is significant at 0.05 level (two-sided)**

WTP = Willingness to pay

CTA = Cross trafic alert

PDC = Park distance control

BSD = Blind spot detection

LDW = Lane departure warning

TPMS = Tire pressure monitoring system

level ($p \leq 0.05$) using the Levene test. In the case of variance heterogeneity, the Welch's t-test was used to determine significant differences. The Pearson correlation coefficient shows the correlation between age and willingness to pay.

Table 3 shows that willingness to pay tends to increase with age. The mean values for all five driver assistance systems are higher in the two older age groups (60–69 years and 70 years and older) than in the younger age group (50–59 years). A detailed evaluation shows differences between the driver assistance systems.

Table 3 Willingness to pay of older drivers for driver assistance systems by age group (two-sample t-test for independent samples, n = 181)

	Group	N	Mean value	SD	T	df	p	Hypothesis
a) Cross traffic alert	50-59	49	319,59	175,05	-,859	123	,392	discarded
	60-69	76	349,14	195,51				
	60-69	76	349,14	195,51	,554	130	,580	discarded
	70-90	56	331,43	160,54				
	50-59	49	319,59	175,05	-,361	103	,719	discarded
	70-90	56	331,43	160,54				
b) Park distance control	50-59	49	300,10	108,91	-2,311	123	,022	Accepted
	60-69	76	365,39	177,17				
	60-69	76	365,39	177,17	,220	130	,826	discarded
	70-90	56	358,39	185,24				
	50-59	49	300,10	108,91	-1,929	103	,053	Accepted
	70-90	56	358,39	185,24				
c) Blind spot detection	50-59	49	269,29	125,57	-2,518	123	,013	Accepted
	60-69	76	341,51	173,52				
	60-69	76	341,51	173,52	-,094	130	,925	discarded
	70-90	56	344,38	171,70				
	50-59	49	269,29	125,57	-2,526	103	,013	Accepted
	70-90	56	344,38	171,70				
d) Lane departure warning	50-59	49	296,33	151,02	-1,931	123	,054	Accepted
	60-69	76	356,38	180,77				
	60-69	76	356,38	180,77	,520	130	,604	discarded
	70-90	56	340,36	166,52				
	50-59	49	296,33	151,02	-1,411	103	,090	discarded
	70-90	56	340,36	166,52				
e) Tire pressure monitoring system	50-59	49	204,08	75,23	-2,723	123	,007	Accepted
	60-69	76	256,58	120,58				
	60-69	76	256,58	120,58	-,565	130	,573	discarded
	70-90	56	270,09	154,06				
	50-59	49	204,08	75,23	-2,727	103	,008	Accepted
	70-90	56	270,09	154,06				

For four of the five surveyed driver assistance systems (park distance control, blind spot detection, lane departure warning, tire pressure monitoring system), hypothesis H_2 (the willingness of older drivers to pay for age-appropriate driver assistance systems increases with the average age of the different age groups) could be confirmed. Table 3, for example, shows significant differences in the statements about the park distance control (Table 3b): the willingness to pay of the older people (60 years and older) is significantly higher than that of the younger ones (50–59 years). Upon closer inspection, it is slightly, but not significantly higher in the group of 60–69 year olds (mean M = 365.39 with SD = 177) than in the age group of 70 year olds and older (M = 358.39 with SD = 185.24), but in both groups it is significantly higher than in the younger age group (M = 300.10; SD = 108.91). The willingness to pay of the older people is also higher than that of the younger ones for the blind spot detection (Table 3c), the lane departure warning (Table 3d), and the tire pressure monitoring system (Table 3e), with the difference that it is highest for the blind spot detection and the tire pressure monitoring system in the age group of 70 year olds and older. For the lane departure warning and the park distance control, the willingness to pay is highest in the 60–69 year age group.

Although the statements also generally apply to the cross-traffic alert (Table 3a), the results are not significant. The age group of 60-69 year old drivers shows the highest willingness to pay (M = 349.14; SD = 195.51), followed by the age group of 70 year olds and older (M = 331.43; SD = 160.54) and the age group of 50–59 year olds (M = 319.59; SD = 175.05).

The survey of the three age groups of older drivers (50–59 years, 60–69 years, 70 years and older) on their willingness to pay thus showed that especially people of higher age are more willing to pay for safety-relevant assistance systems (cross-traffic alert, park distance control, blind spot detection, lane departure warning, and tire pressure monitoring system), significantly more so for four assistance systems. The research hypothesis is therefore clearly confirmed. This result also shows that a differentiation by age groups is fundamentally necessary.

4.2 Results of the Acceptance Examination

This section first summarizes (1) the results of the correlation between age and acceptance of age-appropriate driver assistance systems, then (2) differences in acceptance between the age groups, and finally (3) the influence of acceptance on willingness to pay.

(1) Correlation between Age and Acceptance

The examination of the correlation between age and acceptance of the selected driver assistance systems did not yield significant results. Hypothesis H_3 must therefore be rejected for these driver assistance systems (see Table 4). However, it was found that with experience and product knowledge, the acceptance of age-appropriate driver assistance systems increases, particularly among older drivers. These results also show that there is no clear linear relationship between age and acceptance, thus supporting the need for differentiation according to age groups of older drivers.

When considering the individual factors influencing acceptance, it is evident that perceived usefulness has the greatest impact on usage intention, i.e., the intention to use a specific driver assistance system. Perceived ease of use also has a significant direct influence. If the operation is considered simple, the intention to use increases. Both perceived usefulness and perceived ease of use are significantly influenced by trust in the technology, which also affects acceptance. Personal innovation and subjective norms, on the other hand, only slightly influence usage intention.

Table 4 Correlation table Acceptance (n = 181)

	1. Age	2. Exp.	3. Prod knowl.	4. Net inc.	5. Acc. CTA	6. Acc. PDC	7. Acc. BSD	8. Acc. LDW	9. Acc. TPMS
1. Age	-								
2. Experience	,029	-							
3. Product knowledge	,024	**,708****	-						
4. Net income	-,058	**,241****	**,363****	-					
5. Acc. CTA	,007	**,247****	**,289****	,136	-				
6. Acc. PDC	,112	**,221****	**,296****	**,257****	**,680****	-			
7. Acc. BSD	,135	**,262****	**,326****	,132	**,714****	**,744****	-		
8. Acc. LDW	,041	**,270****	**,308****	**,165***	**,606****	**,657****	**,736****	-	
9. Acc. TPMS	,025	**,239****	**,290****	,145	**,617****	**,589****	**,620****	**,621****	-

* Correlation is significant at 0.01 level (two-sided)

**** Correlation is significant at 0.05 level (two-sided)**

WTP = Willingness to pay
Acc. = Acceptance

CTA = Cross traffic alert
PDC = Park distance control
BSD = Blind spot detection
LDW = Lane departure warning
TPMS = Tire pressure monitong system

(2) Differences in Acceptance between Age Groups

With the help of the mentioned two-sample t-test for independent samples, differences in acceptance between the three age groups (50–59 years, 60–69 years, and 70 years and older) were to be identified, as in the examination of the correlation between age and willingness to pay, with a significance level of $p \leq 0.05$ (highly significant) and $p \leq 0.10$ (slightly significant). Here too, all age groups were compared with each other (Table 5) and the equality of variances at the same significance level ($p \leq 0.05$) was determined with the Levene test. The Pearson correlation coefficient shows the correlation between age and acceptance.

Table 5 shows that acceptance tends to increase with age, the mean values for all five driver assistance systems in the two higher age groups (60–69 years and 70 years and older) are higher than in the younger age group (50–59 years). A detailed evaluation shows differences between the driver assistance systems.

The age group 60–69 years rates all five assistance systems highest. They are also most willing to pay for driver assistance systems. The 60–69 year olds rate the cross-traffic alert (M = 5.73; SD = 0.911, see Table 5a) significantly higher than older people (M = 5.41; SD = 0.958), possibly also because this driver assistance system is less known to the age group and higher than the age group 50–59 years (M = 5.31; SD = 0. 958). Men rate the cross-traffic alert significantly higher (M = 5.58; SD = 0.923) than women (M = 5.29; SD = 0.886).

The parking aid (Table 5b) and the blind spot detection (Table 5c) are rated higher by the 50–59 year olds (M = 5.48; SD = 1.016 and M = 5.28; SD = 1.264) than by the oldest subjects (70 years and older), men again higher than women. Conversely, the lane departure warning (Table 5d) and the tire pressure monitoring system (Table 5e) are rated higher by the 70–90 year olds (M = 5.11; SD = 1.063 and M = 5.43; SD = 0.966) than by the age group of 50–59 year olds. The significantly different results again demonstrate the need for differentiation of older drivers by age groups.

The survey of older drivers in the first main examination confirms Hypothesis H_4 (the acceptance of age-appropriate driver assistance systems *increases* with the average age of the age groups) only for the younger age groups 50 to 69 and for men.

A consideration of the individual factors influencing acceptance shows only a few significant results. This includes that the influence of trust on perceived usefulness is significantly higher for the two older groups of older drivers (60–69 year olds and 70–90 year olds) than for the younger group (50–59 year olds). Furthermore, it is shown that for the group of 50–59 year olds, the influence of people in the environment (the subjective norm) is higher than for the two older groups.

Table 5 Acceptance of the five driver assistance systems by age groups of older drivers (Two-sample t-test for independent samples n = 181)

	Group	N	Mean value	SD	T	df	p	Hypothesis
a) Cross trafic alert	50-59	49	5,31	,968	-2,468	123	,015	Accepted
	60-69	76	5,73	,911				
	60-69	76	5,73	,911	1,949	130	,053	Accepted
	70-90	56	5,41	,958				
	50-59	49	5,31	,968	-,545	103	,587	Discarded
	70-90	56	5,41	,958				
b) Park distance control	50-59	49	5,48	1,016	-2,784	123	,006	Accepted
	60-69	76	5,93	,791				
	60-69	76	5,93	,791	1,021	130	,309	Discarded
	70-90	56	5,78	,897				
	50-59	49	5,48	1,016	-1,614	103	,110	Discarded
	70-90	56	5,78	,897				
c) Blind spot detection	50-59	49	5,28	1,264	-2,886	123	,005	Accepted
	60-69	76	5,84	,907				
	60-69	76	5,84	,907	,916	130	,361	Discarded
	70-90	56	5,70	,938				
	50-59	49	5,28	1,264	-1,913	103	,058	Discarded
	70-90	56	5,70	,938				
d) Lane departure warning	50-59	49	4,99	1,106	-2,496	123	,014	Accepted
	60-69	76	5,44	,890				
	60-69	76	5,44	,890	1,945	130	,054	Accepted
	70-90	56	5,11	1,063				
	50-59	49	4,99	1,106	-,552	103	,582	Discarded
	70-90	56	5,11	1,063				
e) Tire pressure monitoring system	50-59	49	5,33	1,141	-3,005	123	,003	Accepted
	60-69	76	5,88	,878				
	60-69	76	5,88	,878	2,765	130	,007	Accepted
	70-90	56	5,43	,966				
	50-59	49	5,33	1,141	-,477	103	,634	Discarded
	70-90	56	5,43	,966				

(3) Influence of Acceptance on Willingness to Pay

With a structural equation modeling (Fig. 7), the model (Fig. 6) of the influence of acceptance on willingness to pay is examined, which meets all essential validity and reliability criteria. The analysis of the path connection shows a highly significant influence of acceptance on willingness to pay for all surveyed driver assistance systems, thus confirming hypothesis $H_{5.}$

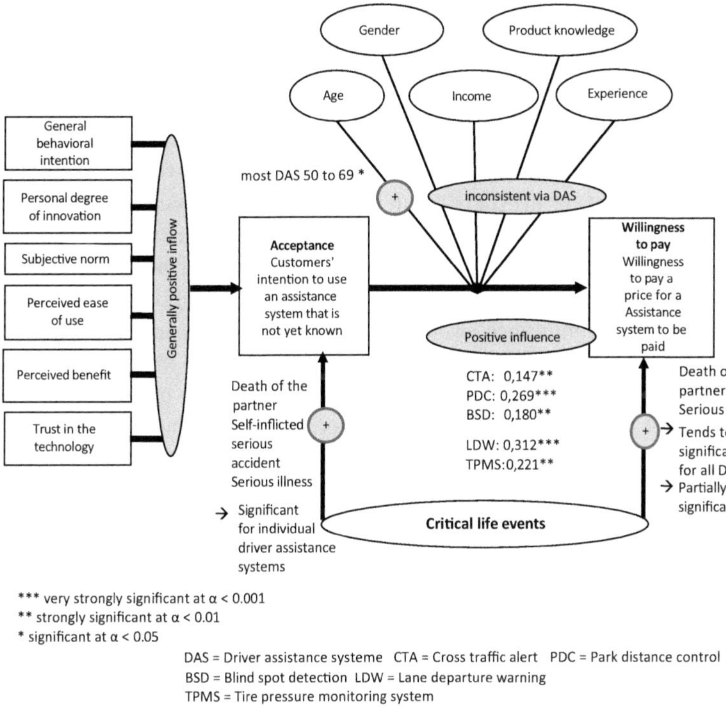

Fig. 7 Results of the Structural equation modelling (n = 181)

However, a multi-group analysis did not provide any evidence of significant differences in the path connection between acceptance and willingness to pay between the study group (181 people aged 50 and older) and the comparison group (46 younger subjects).

The study results show that the acceptance of driver assistance systems, contrary to what was expected in hypothesis H_1, does not clearly correlate with age. This is also confirmed by the examination of the three age groups, according to which - conditionally significant - the acceptance of the 60-69 year olds is highest. It correlates with age in the younger age groups (50-69 years) and positively influences the willingness to pay of all age groups. The remaining moderating variables, e.g. gender and income differences, have no significant influence on the relationship between acceptance and willingness to pay. However, for people who have a higher product knowledge and more experience with driver assistance systems, the influence of acceptance on willingness to pay is slightly increased (see also Fig. 7).

4.3 Examniation of the Influence of Critical Life Events on Acceptance and Willingness to Pay

The two-sample t-test for unpaired samples was also used to test hypothesis H_6 (influence of critical life events on acceptance and willingness to pay for age-appropriate driver assistance systems) at a significance level of $p \leq 0.05$. For this purpose, the subjects of the study group were differentiated according to whether they have experienced presented life events or not (retirement, death of the partner, birth of the first grandchild, self-inflicted serious accident, accident caused by others, serious illness, chronic illness). The equality of variance was determined again at the same significance level ($p \leq 0.05$) with the Levene test. In the presence of variance heterogeneity, the Welch's t-test was also used here. The results of the investigation of the differences (1) of the willingness to pay and then (2) of the acceptance depending on critical life events are now shown.

(1) Differences in Willingness to Pay Depending on Critical Life Events
The results of the examination show that the birth of the first grandchild or chronic diseases have no significant influence on the willingness of older drivers to pay for driver assistance systems. The mean values of willingness to pay remain almost unchanged for all assistance systems.

Retirement as well as self- or other-caused serious accidents do increase the willingness to pay, but not significantly. In contrast, the death of the - possibly helpful on trips - partner is significant. The willingness to pay increases for all driver assistance systems, with the exception of the less known cross-traffic alert. Especially important after the death of the partner appears to be the blind spot detection. Serious illnesses also increase the willingness to pay for all driver assistance systems, significantly for the blind spot detection and the lane departure warning (see also Fig. 7).

(2) Differences in Acceptance Depending on Critical Life Events
The results of the examination show that with the exception of a non-self-inflicted serious accident, all surveyed life events have an influence on the acceptance of driver assistance systems by older drivers.

While retirement or the birth of the first grandchild hardly increase the acceptance of age-appropriate driver assistance systems, the death of the partner, a self-inflicted serious accident, and a serious or chronic illness, on the other hand, have a clear influence on the acceptance of driver assistance systems. The death of the partner, a self-inflicted serious accident, a serious or chronic illness have a particular influence on the acceptance of the cross-traffic alert, a chronic illness also on the acceptance of the park distance control.

The results suggest that critical life events can have an influence on both the willingness to pay and the acceptance of age-appropriate driver assistance systems, even if hypothesis H_5 can only be confirmed conditionally. This is certainly also due to the fact that the samples are relatively small and require larger-scale examniations. However, it should be noted that the death of the partner and a serious illness significantly increase both the willingness to pay and the acceptance of age-appropriate driver assistance systems.

5 Change in Acceptance and Willingness to Pay through Improvement of Driver Assistance Systems for Older Drivers

A comparison of the first and second main examination (in $t_0$and in t_1, cf. Chapter "Older Drivers: Investigation and Sample in the ALFASY Project") shows a clear and significant increase in the willingness to pay of older drivers due to improved (acoustic) driver assistance systems. However, this does not change the acceptance of the individual driver assistance systems. Age also has no influence on acceptance and willingness to pay for improved assistance systems, as the comparison of the study group (50 years and older) with the comparison group of younger drivers shows. Willingness to pay is influenced by other factors.

Differentiated by age, however, the willingness to pay of the 50 to 69 year olds for all driver assistance systems increases due to the improvement of the assistance systems in the second main examination in t_1 (Table 6a), the age group 60–69 years significantly, in the age group 60–69 years with the exception of the lane departure warning (Table 6b). Only for the oldest subjects (70 years and older, Table 6c) the improvement of the driver assistance systems has no significant influence on the willingness to pay.

The increase in willingness to pay due to the improvement of driver assistance systems in the ALFASY project also applies to the comparison group with younger subjects. Their willingness to pay is higher for all improved assistance systems than that of the older age groups, e.g. 552 € for a cross-traffic alert, for which older drivers are only willing to pay an average of 482 €.

The increase in willingness to pay could, however, not only be due to the technical improvement of the acoustic driver assistance systems, but also to the fact that the respondents have gained more experience and product knowledge.

Table 6 Change in willingness to pay due to improvement of driver assistance systems

Age group			Mean value	N	Standard deviation	T	df	sig. (2-sided)	
a) Age group of 50 to 59-year-old drivers	WTP CTA	t0	327,18	35,00	187,346	-3,107	35	,004	Improvement
		t1	493,56	35,00	302,426				
	WTP PDC	t0	295,64	35,00	98,637	-2,784	35	,008	Improvement
		t1	452,79	35,00	373,107				
	WTP BSD	t0	271,41	35,00	125,263	-3,018	35	,005	Improvement
		t1	393,08	35,00	287,185				
	WTP LDW	t0	295,26	35,00	152,056	-2,153	35	,038	Improvement
		t1	386,90	35,00	257,795				
	WTP TPMS	t0	201,92	35,00	61,608	-2,794	35	,008	Improvement
		t1	291,54	35,00	210,760				
b) Age group of 60 to 69 year old drivers	WTP CTA	t0	347,12	65,00	195,167	-3,141	65	,003	Improvement
		t1	458,59	65,00	344,184				
	WTP PDC	t0	367,58	65,00	177,307	-3,306	65	,002	Improvement
		t1	484,33	65,00	338,515				
	WTP BSD	t0	344,92	65,00	177,616	-2,361	65	,021	Improvement
		t1	436,77	65,00	377,781				
	WTP TPMS	t0	251,89	65,00	121,078	-2,665	65	,010	Improvement
		t1	326,83	65,00	256,934				
c) Age group of drivers over 70 years old	WTP CTA	t0	331,20	183,00	171,709	-4,711	182	,000	Improvement
		t1	482,72	183,00	437,335				
	WTP PDC	t0	339,95	183,00	155,593	-5,535	182	,000	Improvement
		t1	466,69	183,00	345,870				
	WTP BSD	t0	328,83	183,00	160,100	-3,822	182	,000	Improvement
		t1	450,87	183,00	478,854				
	WTP LDW	t0	336,12	183,00	165,442	-2,691	182	,008	Improvement
		t1	403,60	183,00	337,344				
	WTP TPMS	t0	244,54	183,00	117,805	-3,339	182	,001	Improvement
		t1	320,33	183,00	312,361				

WTP = Willingness to pay
CTA = Cross traffic alert
PDC = Park distance control
BSD = Blind spot detection
LDW = Lane departure warning
TPMS = tire pressure monitoring system

However, as willingness to pay does not increase with increasing experience and product knowledge, the higher willingness to pay is largely due to improved assistance systems. This result makes it clear that driver assistance systems would need to be developed even more specifically for older people in the future and adapted to their needs.

The departure from studies in the driving simulator and the testing of the improved (acoustic) driver assistance systems in the ALFASY project on a test drive in the third main examination (in t_2) confirms the increase in average willingness to pay and the unchanged acceptance of older drivers compared to the first main examination (in t_0). Because the sample for the test drive had to be reassembled (cf. Chapter "Older Drivers: Investigation and Sample in the ALFASY Project" in this book), the samples are not comparable and the results can only be understood as trend statements.

6 Market Potentials of Age-Appropriate Driver Assistance Systems for Older Drivers - Conclusions for Further Research

The business economic examination of willingness to pay and acceptance of age-appropriate driver assistance systems in the ALFASY project shows that they are particularly positively evaluated by the age groups 50 to end of 60, who are also willing to pay for them.

Willingness to pay allows an estimation of the revenue potentials and thus the market potentials on the "silver market" (here persons from 50) with about 36.8 million people of this age group in Germany (Statistisches Bundesamt 2019). To tap into this, the market for age-appropriate driver assistance systems must be differentiated by market segments. The investigation results on important customer wishes, acceptance and critical life events such as death of the partner and serious illnesses, which influence the willingness to pay, provide indications for a targeted market approach and the design of the benefit promise as an element of business models. The investigations in the ALFASY project were supplemented by questions about the customer-specific benefit of age-appropriate driver assistance systems. They support the design of a benefit promise.

Even if it is currently not yet foreseeable when autonomous driving vehicles will be offered, the worldwide market introduction, substitution and diffusion will still take some time, age-appropriate driver assistance systems significantly improve comfort and safety and enable older people to remain mobile for longer. For manufacturers and suppliers, the "silver market" offers great potentials.

References

Agarwal, R., & Prasad, J. (1998). A conceptual and operational definition of personal innovativeness in the domain of information technology. *Information Systems Research, 9*(2), 204–215.

Ajzen, I. (1991). The theory of planned behavior. *Organizational Behavior and Human Decision Processes, 50*(2), 179–211.

Ajzen, I., & Fishbein M. (1980). *Understanding attitudes and predicting social behavior.* Englewood Cliffs, Prentice-Hall.

BCG (The Boston Consulting Group), & MEMA (The Motor & Equipment Manufacturers Association) (2015). A Roadmap to safer driving through advanced driver assistance systems. *Auto Tech Review, 5*, 20–25.

Blundell, R., Browning, M., & Meghir, C. (1994). Consumer demand and the life-cycle allocation of household expenditures. *The Review of Economic Studies, 61*(1), 57–80.

Blythe, P. T., & Curtis, A. M. (2004). Advanced driver assistance systems: Gimmick or reality. *Conference Proceedings at the 11th World Congress on Intelligent Transport Systems and Services, Nagoya.*

Boot, W., & Scialfa, C. (2016). The aging road user and technology to promote safe mobility for life. In: S. Kwon (Hrsg.), *Gerontechnology: Research, Practice, and Principles in the Field of Technology and Aging* (S. 207–222), New York, Springer.

Braun, H., Gärtner, M., Trösterer, S., Akkermans, L. E., Seinen, M., Meschtscherjakov, A., & Tscheligi, M. (2019). Advanced driver assistance systems for aging drivers: Insights on 65+ drivers' acceptance of and intention to use ADAS. *Proceedings of the 11th International Conference on Automotive User Interfaces and Interactive Vehicular Applications*, Utrecht, 123–133.

Burghard, E. (2005). *Fahrkompetenz im Alter – Die Aussagekraft diagnostischer Instrumente bei Senioren und neurologischen Patienten.* Dissertation, Ludwig-Maximilian-Universität München.

Chatterjee, K., & Scheiner, J. (2015). *Understanding changing travel behaviour over the life course: Contributions from biographical research.* A resource paper for the Workshop "Life-Oriented Approach for Transportation Studies". Presented at the 14th International Conference on Travel Behaviour Research (IATBR), Windsor.

Clark, B., Chatterjee, K., & Melia, S. (2016). Changes in level of household car ownership: the role of life events and spatial context. *Transportation*, *43*(4), 565–599.

Davidse, R. J. (2006). Older drivers and ADAS: Which systems improve road safety? *IATSS research*, *30*(1), 6–20.

Davis, F. D, Bagozzi, R. P., & Warshaw, P. R. (1989). User acceptance of computer technology: A comparison of two theoretical models. *Journal of Management Science*, *35*(8), 982–1003.

Dudenhöffer, K. (2015). *Akzeptanz von Elektroautos in Deutschland und China: Eine Untersuchung von Nutzungsintentionen im Anfangsstadium der Innovationsdiffusion.* Dresden, Springer.

Edwards, J. D., Perkins, M., Ross, L. A., & Reynolds, S. L. (2009). Driving status and three-year mortality among community-dwelling older adults. *The Journals of Gerontology: Series A*, *64*(2), 300–305.

Elder, G.H. (1994). Time, human agency and social change: Perspectives on the life course. *Social Psycology Quarterly*, *57*(1), 4–15.

Engeln, A. (2001). *Aktivität und Mobilität im Alternsprozess*. Aachen, Shaker.

Fagan, M. H., Neill, S., & Wooldridge, B. R. (2008). Exploring the intention to use computers: An empirical investigation of the role of intrinsic motivation, extrinsic motivation and perceived ease of use. *Journal of Computer Information Systems*, *48*(3), 31–37.

Fernández-Villaverde, J., & Krueger, D. (2007). Consumption over the life cycle: Facts from consumer expenditure survey data. *The Review of Economics and Statistics*, *89*(3), 552–565.

Filipp, S. H. (1995). *Kritische Lebensereignisse*. 3. Aufl., Weinheim: Psychologie Verlags Union.

Fuchs-Heinritz, W. (2009). *Biographische Forschung: Eine Einführung in Praxis und Methoden*. Wiesbaden: VS Verlag für Sozialwissenschaften.

Glenn, N. D. (2003). Distinguishing age, period, and cohort effects. In J. T. Mortimer & M. J. Shanahan (Hrsg.), *Handbook of the life course* (S. 465–476). Boston, Springer.

Günthner, T., Proff, H., Jovic, J., & Zeymer, L. (2020). Tapping into market opportunities in aging societies – the example of advanced driver assistance systems in the transition to autonomous driving. *International Journal of Automotive Technology and Management*, im Druck.

Holley-Moore, G., & Creighton, H., (2015). *The future of transport in an ageing society*. International London: Longevity Centre-UK.

Horlacher, K. D. (2000). Kritische Lebensereignisse. In: M. Amelang (Hrsg.), *Determinanten individueller Unterschiede. Enzyklopädie der Psychologie* (S. 455–486). Göttingen: Hogrefe.

Karthaus, M., Falkenstein, M., & Toepper, M. (2016). Functional changes and driving performance in older drivers: Assessment and Interventions. *Geriatrics*, *1*(12), 1–18.

Kaur, K., & Rampersad, G. (2018). Trust in driverless cars: Investigating key factors influencing the adoption of driverless cars. *Journal of Engineering and Technology Management, 48*, 87–96.

Kloss, D., & Kunter, M. (2016). The van Westendorp price-sensitivity meter as a direct measure of willingness-to-pay. *European Journal of Management*, *16*(2), 45–54.

Kwon, O., Choi, K., & Kim, M. (2007). User acceptance of context-aware services: Self-efficacy, user innovativeness and perceived sensitivity on contextual pressure. *Journal of Behavior & Information Technology*, *26*(6), 483–498.

König, M., & Neumayr, L. (2017). Users' resistance towards radical innovations: The case of the self-driving car. *Transportation Research Part F*, *44*, 42–52.

Leifer, R., Corarelli O'Conner, D., & Rice, M. (2001). Implementing radical innovations in mature firms: The role of hubs. *Academy of Management Executive*, *15*(3), 102–113.

Mayer, R. C., Davis, J. H., & Schoorman, F. D. (1995). An integrative model of organizational trust. *Academy of Management Review*, *20*(3), 709–734.

Midgley, D. F., & Dowling, G. R. (1978). Innovativeness: The concept and its measurement. *Journal of Consumer Research*, *4*(4), 229–242.

Miles, D. (1997). A household level study of the determinants of income and consumption. *The Economic Journal*, *107*(440), 1–25.

Miller, K. M., Hofstetter, R., Krohmer, H., & Zhang, Z. J. (2011). How should consumers' willingness to pay be measured? An empirical comparison of state-of-the-art approaches. *Journal of Marketing Research*, *48*(1), 172–184.

Montada, L. (2008). Grundlagen der Entwicklungspsychologie. *Entwicklungspsychologie*, *6*, 3–48.

Moschis, G. P. (2012). Consumer behavior in later life: Current knowledge, issues, and new directions for research. *Psychology & Marketing*, *29*(2), 57–75.

Musselwhite, C., & Haddad, H. (2018). Older people's travel and mobility needs: a reflection of a hierarchical model 10 years on. *Quality in Ageing and Older Adults*, *19*(2), 87–105.

Musselwhite, C., Holland, C., & Walker, I. (2015). The role of transport and mobility in the health of older people. *Journal of Transport & Health*, *2*(1), 1–4.

Müggenburg, H., & Lanzendorf, M. (2015). Beruf und Mobilität – Eine intergenerationale Untersuchung zum Einfluss beruflicher Lebensereignisse auf das Verkehrshandeln. In J. Scheiner & C. Holz-Rau (Hrsg.), *Räumliche Mobilität und Lebenslauf* (S. 79–95). Wiesbaden: Springer.

Nobis, C., Kuhnimhof, T., Follmer, R., & Bäumer, M. (2019). Mobilität in Deutschland – Zeitreihenbericht 2002 – 2008 – 2017. *Studie von infas, DLR, IVT und infas 360 im Auftrag des Bundesministeriums für Verkehr und digitale Infrastruktur,* FE-Nr. 70.904/15.

Piao, J., McDonald, M., Henry, A., Vaa, T., & Tveit, O. (2005). An assessment of user acceptance of intelligent speed adaptation systems. *Proceedings to 8th IEEE Intelligent Transportation Systems,* Vienna, 1045-1049.

Proff, H. (2019). *Multinationale Automobilunternehmen in Zeiten des Umbruchs: Herausforderungen – Geschäftsmodelle – Steuerung.* Wiesbaden: Springer.

Proff, H., & Fojcik, T. M. (2014). Accelerating market diffusion of battery electric vehicles through alternative mobility concepts. *International Journal of Automotive Technology and Management, 14*(3), 347–368.

Proff, H., Szybisty, G., Fojcik, T.M., Cremer, C. (2018). Neue Geschäftsmodelle für Dienstleistungsinnovationen im Automobilhandel für die Elektromobilität. In: H. Proff, M. Borchert & G. Schmitz (Hrsg.), *Dienstleistungsinnovationen und Elektromobilität: Der Automobilhandel als ganzheitlicher Lösungsanbieter* (S. 5–80). Wiesbaden: Springer-Gabler.

Roll, O., Pastuch, K., & Buchwald, G. (2018). *Praxishandbuch Preismanagement: Strategien-Management-Lösungen.* Weinheim: John Wiley & Sons.

Schleiffer, N. (2024). *Ausgestaltung des Nutzenversprechens innovativer Produkte der Automobilindustrie zur Verbesserung des wahrgenommenen Nutzens der Zielkunden weltweit: Eine empirische Untersuchung zum Übergang in die Elektromobilität in Deutschland und China.* Wiesbaden: SpringerGabler.

Schulz, R., Beach, S. R., Matthews, J. T., Courtney, K., Dabbs, A. D., Mecca, L. P., & Sankey, S. S. (2013). Willingness to pay for quality of life technologies to enhance independent functioning among baby boomers and the elderly adults. *The Gerontologist, 54*(3), 363–374.

Simon, H., &Fassnacht, M. (2016). *Preismanagement: Strategie-Analyse-Entscheidung-Umsetzung.* Wiesbaden: Springer.

Son, J., Park, M., & Park, B. B. (2015). The effect of age, gender and roadway environment on the acceptance and effectiveness of Advanced Driver Assistance Systems. *Transportation Research Part F: Traffic Psychology and Behaviour, 31*, 12–24.

Souders, D. J., Best, R., & Charness, N. (2017). Valuation of active blind spot detection systems by younger and older adults. *Accidental Analysis and Prevention, 106*, 505–514.

Souders, D. J., & Charness, N. (2016). Challenges of older drivers' adoption of advanced driver assistance systems and autonomous vehicles. *Human Aspects of IT for the Aged Population. Healthy and Active Aging*, 428–440.

Statistisches Bundesamt (2019). Bevölkerung in Deutschland. Verfügbar unter https://service.destatis.de/bevoelkerungspyramide/index.html#!y=2018&a=50,100&v=2&g

Stevens, S. (2012). The relationship between driver acceptance and system effectiveness in car-based collision warning systems: Evidence of an overreliance effect in older drivers? *SAE International Journal of Passenger Cars – Electronic and Electrical Sys*tems, *5*(1), 114–124.

Trübswetter, N. (2015). *Akzeptanzkriterien und Nutzungsbarrieren älterer Autofahrer im Umgang mit Fahrerassistenzsystemen.* Dissertation, Technische Universität München.

Trübswetter, N., & Bengler, K. (2011). Systematische Modellierung des zukünftigen Unterstützungspotentials im Straßenverkehr. In Gesellschaft für Arbeitswissenschaft (Hrsg.), *Mensch, Technik, Organisation – Vernetzung im Produktentstehungs- und -herstellungsprozess* (S. 841–844). Dortmund: GfA-Press.

Uteng, T. P., Julsrud, T. E., & George, C. (2019): The role of life events and context in type of car share uptake: Comparing users of peer-to-peer and cooperative programs in Oslo. *Transportation Research Part D: Transport and Environment, 71*, 186–206.

Van Westendorp, P. H. (1976). NSS Price Sensitivity Meter (PSM) – A new approach to study consumer perception of prices. *Proceedings of the 29th ESOMAR Congress,* Venice, 139–167.

Varian, H. R. (2014). *Grundzüge der Mikroökonomik*. 8. Aufl., Oldenburg: München.

Venkatesh, V., & Bala, H. (2008). Technology acceptance model 3 and a research agenda on interventions. *Decision Science*, *39*(2), 273–315.

Venkatesh, V., & Davis, F. D. (2000): A theoretical extension of the technology acceptance model: four longitudinal field studies. *Management Science*, *46*(2), 186–204.

Viborg, N. (1999). Older and younger driver's attitudes toward in-car ITS a questionnaire survey. *Bulletin 181*. Lund, Sweden: Department of Technology and Society, Lund Institute of Technology.

Winner, H., & Schopper, M. (2015): Adaptive cruise control. In: H. Winner, S. Hakuli, G. Wolf (Hrsg.), *Handbuch Fahrerassistenzsysteme – Grundlagen, Komponenten und Systeme für aktive Sicherheit und Komfort* (S. 851–891). Wiesbaden: Vieweg+Teubner.

Yang, F. (2009). Consumption over the life cycle: How different is housing, *Review of Economic Dynamics*, *12*(3), 423–443.

Age-Appropriate Driver Assistance Systems: Technical, Psychological and Business Aspects—a Summary

Heike Proff, Matthias Brand and Dieter Schramm

Contents

It is the task of science and industry to increase the objective safety in road traffic for all road users through appropriate technical solutions and assistance systems (see Chapter "Mobility in Old Age—An Introduction"), because

- The proportion of older people in more developed countries will continue to increase (e.g., Statistisches Bundesamt 2019),

Prof. Dr. Heike Proff, Prof. Dr. Matthias Brand, Prof. Dr.-Ing. Dieter Schramm, all University of Duisburg-Essen.

H. Proff (✉)
Lehrstuhl für ABWL & Internationales Automobilmanagement,
Universität Duisburg-Essen, Duisburg, Germany
e-mail: heike.proff@uni-due.de

M. Brand
Allgemeine Psychologie: Kognition, Universität Duisburg-Essen, Duisburg, Germany
e-mail: matthias.brand@uni-due.de

D. Schramm
Lehrstuhl für Mechatronik, Universität Duisburg-Essen, Duisburg, Germany
e-mail: dieter.schramm@uni-due.de

H. Proff et al. (eds.), *Age-appropriate Driver Assistance Systems*,
https://doi.org/10.1007/978-3-658-44838-7_10

- More and more older people possess a driver's license (Kraftfahrt-Bundesamt 2018), as a car means independence, participation, activity, and thus quality of life (e.g., Engeln 2001; Burghard 2005),
- Older people often have problems in road traffic due to age-related declines in cognitive, visual, and motor skills (Brand and Markowitsch 2010; MoPact 2014) and
- The age group of over 50-year-olds is able and mostly also willing to pay a premium for products that increase safety (Wild 2014).

Suitable assistance systems enable older drivers to compensate for their poorer information intake and processing compared to younger drivers, and to actively participate in motorized individual traffic. Inaddition, a contribution is also made to the safety of all road users (Kubitzki 2013).

Although technologies and highly developed driver assistance systems that support both older and younger drivers are available and are constantly being improved, the sales of driver assistance systems still fall far short of the expectations of car manufacturers (e.g., Wild 2014; Winner and Schopper 2015).

The aim of the ALFASY project (**Al**tersgerechte **Fa**hrerassistenz**sy**steme, i.e. Age-appropriate driver assistence systems) was therefore to develop, build and test an acoustic human-machine interface for selected driver assistance systems that is directed at the needs of the steadily growing group of older drivers (50 years and older) and also fits economically into the product portfolio of car manufacturers and suppliers.

This book addressed important technical, psychological, and business aspects of the human-machine interface and the driver assistance systems considered, which were developed in collaboration with three chairs at the University of Duisburg-Essen (Mechatronics, General Psychology/Cognition, General Business Administration & International Automotive Management) with Ford, HEAD acoustics, and the Allround Team. The main results are briefly summarized:

1. Driver assistance systems with an optimized human-machine interface were tested in the ALFASY project by older car drivers compared to a comparison group of younger drivers: first multiple times in the driving simulator, finally in a vehicle on a test track. Due to the high age of some subjects, a query e.g., acute cardiovascular problems and psychological problems was necessary, because people with such problems cannot participate in the tests. Since empirical tests in the driving simulator can be expected to result in so-called simulator sickness, which can lead to a termination of the tests (Chapter "Neuropsychological and

-Physiological Correlates of Older Drivers' Driving Behavior Within Simulated Environments"), reactions in the driving simulator were also tested in the preliminary examination (see Chapter "Driving Behavior of Older People in the Driving Simulator" on the driving behavior of older people). Of a total of 465 older car drivers invited to the tests, 439 people (of which 381 were older than 50 years) participated in the preliminary examination. Physical or psychological restrictions were not apparent, the test drive in the driving simulator was inconspicuous.

2. The various empirical examinations in the ALFASY project show peculiarities of older drivers compared to younger drivers. They reveal a large market potential (see Chappter "Older Drivers: Investigation and Sample in the Alfasy Project"):

- The proportion of men interested in driver assistance systems is significantly higher and increases with age. Female test subjects express more uncertainty about driver assistance systems that are often unknown or less familiar to them.
- The net income of older drivers is higher and increases with age in the preliminary examination (largest sample).
- Daily driving performance decreases with increasing age.
- Older drivers express a latent and open need for driver assistance systems, although less than would be expected considering anticipated physical and psychological limitations. They are particularly looking for support in terms of visibility and perception of the vehicle environment and parking.
- Knowledge of driver assistance systems is rather low and decreases with increasing age.
- However, experiences are generally positive. Older drivers see clear advantages of driver assistance systems, even if they have not yet had their own experiences.
- Older drivers are less willing than younger drivers to rely on technology and delegate decisions, which is not surprising given their limited experience with driver assistance systems.
- Specific wishes for driver assistance systems are rarely mentioned. This is explained by the fact that they are either not known at all or only slightly known.
- For older drivers, low costs are less important than for younger ones.

3. In the ALFASY project, five driver assistance systems were tested: cross-traffic alert, park distance control, lane departure warning, blind spot detection, and

tire pressure monitoring system (Chapter "Driver Assistance Systems - An Overview"). They were made perceptible with acoustic signals to scientifically investigate their benefit for older people.

4. Studies on driver assistance systems in the context of age-appropriate design of user interfaces (Human Machine Interfaces, HMI, Chapter "Driver Assistance Systems in the Context of Age-Appropriate HMI Design") show that offered driver assistance systems at least offer a great support potential for older drivers and increase road safety when supported by age-appropriate user interfaces. If they take into account sensory, physical, and cognitive peculiarities of this target group, fears, reservations, or an underestimation of the benefits of driver assistance systems can be reduced. In Chapter "Driver Assistance Systems in the Context of Age-Appropriate HMI Design", possibilities were shown on how user interfaces can be designed age-appropriately (e.g., larger windshields or intelligent avatars) to better utilize the potential of driver assistance systems for the target group of older people and enable safe mobility.

5. Based on the evaluation of acoustic driver assistance systems for older drivers (Chapter "Evaluation of the Performance of Acoustic Assistance Systems for Older Drivers") with higher reaction speed than to visual and haptic stimuli (e.g., Aditya et al. 2015), a localization system for evaluating these assistance systems with a specially developed microphone arrays (interaction of several microphones) and a neural network was presented. The microphone array records components of the sound field in the vicinity of the ears, which humans also use for orientation or localization. Signals obtained in this way are analyzed by a neural network, which determines the origin position with a resolution of 5°. It was shown that neural networks are significantly more precise compared to conventional beamforming algorithms. The method can be directly transferred to other user groups or applications by exchanging the training and validation data.

6. The driving behavior of older drivers in the driving simulator is documented in Chapter "Driving Behavior of Older People in the Driving Simulator". The setup of the simulator, the experiments carried out, and the effects of simulator sickness (kinetosis) in simulator studies are described and the test results briefly summarized. The study in the ALFASY project supports assumptions that simulator sickness is directly related to acclimatization to such systems, as the symptoms decrease when multiple test drives have been made. The occurrence and intensity of simulator sickness can be reduced by training or acclimatization. The Index of Performance (IOP) can be used to calculate driving performance (Joshi

et al. 2017). The index is composed of several individual ratings such as driving comfort and control activity. The results show a habituation effect in almost all participants of the third test drive with assistance systems compared to the second drive. On average, the parking time decreases, the driven distance becomes longer—with and without changed sound settings.

7. In the ALFASY project, after the tests in the driving simulator, the third main examination was conducted in a vehicle to test acoustic driver assistance systems under real conditions. In Chapter "Implementation of Real Driving Studies - Test Vehicle and Experimental Setup in t2" the experimental setup and systems for higher and lower speed ranges are described and the results summarized. Three acoustic variants were tested:

- non-directional (sounds only come from one speaker, regardless of the position of the obstacle),
- static directional (sounds only come from a speaker facing the obstacle) and
- dynamic directional (only for cross-traffic alert: when the obstacle changes, sounds move across the speakers).

For both the blind spot detection and the lane departure warning, the static directional acoustic variant performs better than the dynamic directional variant and even more clearly than the non-directional variant. This result can be explained by the fact that the non-directional variant provides the least information value (no encoding of the location, i.e. the obstacle or the lane left). When comparing the two directional variants, it was emphasized that the static directional variant already contains enough information about the location. An additional dynamic of the sounds thus brings no further advantage and is perceived as rather disturbing or irritating. Driver assistance systems therefore benefit most from the implementation of a static directional acoustic variant.

8. The neuropsychological and -physiological results can be summarized as follows:

- There was a negative correlation between attention performance and the time needed to adapt to the simulator, from which it can be concluded that higher attention performance accelerates adaptation to a driving simulator.
- Female participants showed a significantly higher rate of occurrence of simulator sickness symptoms compared to men.

- Simulator sickness was less likely to occur in people who have adapted to the simulator.
- The results show no significant correlation between driving performance and stress level before driving or previous experiences with simulator sickness.
- A correlation between age and driving performance was demonstrated in that older participants showed poorer driving performance.
- For predicting driving performance, context-specific tasks seem to be more suitable than classic neuropsychological test procedures.
- There was no correlation between the experience with symptoms of simulator sickness and the subsequent driving performance in the simulator.

9. The investigation of willingness to pay and acceptance of age-appropriate driver assistance systems to estimate the market potentials of older drivers (Chapter "Market Potentials of Older Drivers - Willingness to Pay and Acceptance of Age-Appropriate Driver Assistance Systems") in the first two main examinations of the ALFASY project in the driving simulator show:

- Age and willingness to pay of older drivers correlate only to a limited extent. The correlation is significant for the blind spot detection and tire pressure monitoring system, not significant for the cross-traffic alert, park distance control, and lane depature warning. Of the control variables, only income has a significant influence on the willingness to pay, not experience and product knowledge. Willingness to pay increases with increasing income.
- Differentiation by age groups, however, shows that for four of the five queried driver assistance systems (park distance control, blind spot detection, lane departure warning, tire pressure monitoring system) the hypothesis could be confirmed that the willingness of older drivers to pay for driver assistance systems increases with the average age of the age groups. Especially the age group 50 years to end of 60 is willing to pay for it.
- The investigation of the correlation between age and acceptance of the selected driver assistance systems, on the other hand, did not yield any significant results. However, it was found that with experience and product knowledge, the acceptance of driver assistance systems increases, especially among older drivers.
- Differentiation by age groups of older drivers, however, confirms at least for younger people between 50 to 69 years and for men that the acceptance of driver assistance systems increases with the average age of the age groups.
- A path analysis shows a highly significant acceptance of all queried assistance systems on the willingness to pay.

- There are indications that critical life events can have an influence on both the willingness to pay and the acceptance of age-appropriate driver assistance systems. Especially the death of the partner and a serious illness significantly increase both the willingness to pay and the acceptance of age-appropriate driver assistance systems.

Through the collaboration of all involved partners from science and industry, an acoustic driver assistance system was developed, built, and tested multiple times in the ALFASY project. In this process, the needs of the steadily growing group of older drivers were examined in detail. Statements about the willingness to pay of this customer group allow an estimation of the market potential. The attitudes and behavior of older drivers are difficult to capture because they do not change linearly with age during critical life events. However, the project showed that older drivers form a very attractive market segment. From the survey of a large number of test subjects, many new insights were gained and new research topics were developed.

References

Aditya, J., Bansal, R., Kumar, A., & Singh, K. (2015). A comparative study of visual and auditory reaction times on the basis of gender and physical activity levels of medical first year students. *International Journal of Applied and Basic Medical Research, 5*(2), 124–127.

Brand, M., & Markowitsch, H. J. (2010). Aging and decision-making: a neurocognitive perspective. *Gerontology*, *56*(3), 319–324.

Burghard, E. (2005). Fahrkompetenz im Alter – Die Aussagekraft diagnostischer Instrumente bei Senioren und neurologischen Patienten (Dissertation, Ludwig-Maximilian-Universität München). Digitale Hochschulschriften der Ludwig-Maximilian-Universität München, München.

Engeln, A. (2001). Aktivität und Mobilität im Alternsprozess. Aachen: Shaker.

Joshi, S. S., Maas, N., & Schramm, D. (2017). A vehicle dynamics based algorithm for driver evaluation. In *11th International Conference on intelligent Systems and Control (ISCO)*, p. 5, Karpagam College of Engineering Myleripalayam, Coimbatore, India.

Kraftfahrt-Bundesamt (2018). Bestand an allgemeinen Fahrerlaubnissen im ZFER am 1. Januar 2018 nach Geschlecht, Lebensalter und Fahrerlaubnisklassen. Verfügbar unter https://www.kba.de/DE/Statistik/Kraftfahrer/Fahrerlaubnisse/Fahrerlaubnisbestand/2018/2018_fe_b_geschlecht_alter_fahrerlaubniskl.html?nn=2218648

Kubitzki, J. (2013). Sicherheit und Mobilität älterer Verkehrsteilnehmer. Fachtagung Mobilität im Alter, LandesSeniorenVertretung Bayern e. V., Fürth, 29.04.2013.

MoPact (2014). Mobilising the potential of active ageing in Europe. Verfügbar unter http://mopact.group.shef.ac.uk/

Statistisches Bundesamt (2019). Bevölkerung in Deutschland. Verfügbar unter https://service.destatis.de/bevoelkerungspyramide/index.html#!y=2018&a=50,100&v=2&g
Wild, A. (2014). Zwischen Wunsch und Wirklichkeit: Fahrerassistenzsysteme für ältere Autofahrer. *Zeitschrift für die gesamte Wertschöpfungskette Automobil*, *17*(1), 58–63.
Winner, H., & Schopper, M. (2015). Adaptive cruise control. In H. Winner, S. Hakuli & G. Wolf (Eds.), *Handbuch Fahrerassistenzsysteme – Grundlagen, Komponenten und Systeme für aktive Sicherheit und Komfort* (pp. 851–891). Wiesbaden: Vieweg+Teubner.

GPSR Compliance

The European Union's (EU) General Product Safety Regulation (GPSR) is a set of rules that requires consumer products to be safe and our obligations to ensure this.

If you have any concerns about our products, you can contact us on ProductSafety@springernature.com

In case Publisher is established outside the EU, the EU authorized representative is:

Springer Nature Customer Service Center GmbH
Europaplatz 3
69115 Heidelberg, Germany

Batch number: 10732049

Printed by Printforce, the Netherlands